T0226093

Environmental Footprints and Eco-design of Products and Processes

Series editor

Subramanian Senthilkannan Muthu, SGS Hong Kong Limited, Hong Kong, Hong Kong SAR

More information about this series at http://www.springer.com/series/13340

Miguel Angel Gardetti
Subramanian Senthilkannan Muthu
Editors

Sustainable Luxury, Entrepreneurship, and Innovation

 Springer

Editors
Miguel Angel Gardetti
Center for Studies on Sustainable Luxury
Buenos Aires
Argentina

Subramanian Senthilkannan Muthu
Bestseller
Hong Kong
Hong Kong

ISSN 2345-7651   ISSN 2345-766X   (electronic)
Environmental Footprints and Eco-design of Products and Processes
ISBN 978-981-13-4937-9   ISBN 978-981-10-6716-7   (eBook)
https://doi.org/10.1007/978-981-10-6716-7

Printed on acid-free paper

This Springer imprint is published by Springer Nature
The registered company is Springer Nature Singapore Pte Ltd.
The registered company address is: 152 Beach Road, #21-01/04 Gateway East, Singapore 189721, Singapore

Preface

Luxury depends on cultural, economic, or regional contexts which transform luxury into an ambiguous concept. Also, according to Ricca and Robins (2012) luxury is a source of inspiration, controversy, admiration, and considerable economic success. And, along this controversial line, already in 1999 Robert H. Frank stated in his book "Luxury Fever—Weighing the Cost of Excess" the need to minimize the culture of "excess" to restore the true values of life. And this is in line with the World Commission on Environment and Development (WCED 1987) report, Our Common Future, also known as the Brundtland Report, which defines sustainable development as the development model that allows us to meet present needs, without compromising the ability of future generations to meet their own needs. Luxury is becoming more about helping people to express their deepest values. So, sustainable luxury would not only be the vehicle for more respect for the environment and social development, but it will also be synonym of culture, art, and innovation of different nationalities, maintaining the legacy of local craftsmanship (Gardetti 2011).

Innovation provides the means to create a new reality. It involves converting knowledge, learning, capabilities, and insights into value and creative new perspectives, products, productive outcomes, and business models. Innovation is more than change: It is making incremental and/or radical improvements to systems, technologies, products, processes, and practices (business models) (Rainey 2006).

On the other hand, there are people with a profound perspective toward environmental and social issues and who are well motivated to "break" the rules and promote disruptive solutions to these issues, and most of them are entrepreneurs. These individuals have a number of different roles to play in entrepreneurship or intrapreneurship and innovation, from the imaginative act of setting up a new venture. This involves cognitive and motivational characteristics. In addition, to achieve a profound social change, the role of the personal values is very important: Idealistic values regarding environmental and social goals can be translated into value economic assets (Dixon and Clifford 2007). They have a transformational leadership behavior, inspiring and guiding the fundamental transformation that sustainability requires (Egri and Herman 2000).

The Book

The book begins with a paper by Silvia Ranfagni and Simone Guercini—"The Face of Culturally Sustainable Luxury: Some Emerging Traits from a Case Study." This chapter investigates cultural sustainability in luxury. Seen as a path followed by luxury brands to increase the perception of sustainable luxury, the authors have identified its drivers, prerequisites, and managerial implications. In doing so, they have studied the integration of cultural heritage and luxury through the representative case of a luxury company which produces fabrics expressing local traditions and uses them in its collections.

In the next chapter, "How the Business Model Could Increase the Competitiveness of a Luxury Company?" by Elisa Giacosa, the author verifies how the business model could increase the competitiveness of a luxury company, using a qualitative approach; in particular, considering large- and medium-sized luxury companies which are internationally recognized with highly innovative and entrepreneurial business approach over generations.

Following "Appreciate Mentoring as an Innovative Micro-Practice of Employee Engagement for Sustainability: A Luxury Hotel's Entrepreneurial Journey" was developed by Gulen Hashmi. The focus of this article is to showcase how, within a strong organizational culture of sustainability, appreciative mentoring can be used as an innovative entrepreneurial micro-practice to facilitate positive conversations and engagement in the organizational transformation process of enhancing employee engagement.

Moving on, Miguel Angel Gardetti developed "Entrepreneurship, Innovation and Luxury: The Case of ANTHYIA." This case analyzes the company under the model of sustainable value creation developed by Prof. Stuart L. Hart (Hart 1995, 2005, 2007; Hart and Milstein 1999, 2003) that integrates four aspects: environment, innovation, stakeholder management, and potential for growth. Anthyia is the combination of Ying's personal values and the eastern perspective of ramie in a continuous creativity and innovation process. This is essential for the company's long-term existence and, since it is a dynamic concept, it requires managerial and organizational leadership that are typical of both Ying and Anthyia.

The chapter called "The Communication of Sustainability by Italian Fashion Luxury Brands: A Framework to Qualitatively Evaluate Innovation and Integration " was written by Mosca, Civera, and Casalegno. The chapter has twofold aims and uses a mixed methodological approach. Firstly, it aims at proposing a framework for qualitatively measuring the communication of CSR in an innovative strategic way. Secondly, it seeks to preliminarily investigate the extent of strategic CSR communication among a sample of 30 Italian fashion luxury players, through the content analysis of the CSR communications spread online, which is nowadays a trend that luxury market needs to constantly face.

In turn, Marius Schemken and Benjamin Berghaus in their paper "The Relevance of Sustainability in Luxury from the Millennials' Point of View" investigate, by means of a focus group interview with millennials, in which respect the image of luxury firms from the millennial generation's point of view is influenced by sustainability efforts. The paper concludes that millennials have manifold and ambivalent associations with sustainability and luxury, that apportioned synergies between sustainability and luxury are hardly captured by millennials, and that millennials discuss the matter differently based on their perspective.

The next chapter "Opal Entrepreneurship: Indigenous Integration of Sustainable Luxury in Coober Pedy" was developed by Annette Condello. In a world of diminishing resources, the opal has become a sign of mineral exclusivity for the consumer luxury market and its value as a luxury object comes from gemstone cognoscenti. According to one Australian Aboriginal legend, rainbow-hued opals are believed by some to stir emotions of loyalty and connection to the earth. Regarding the integral indigenous connection of Australia's national gemstone, rarely has one has looked at the spaces where opal veins were once quarried in remote regions in terms of sustainable luxury.

Turning to adaptive reuse and indigenous culture in Coober Pedy—South Australia's opal mining industry—this chapter addresses the existing underground passages as the recyclable integration of a former mining site. In tracking the way in which the community and its rural groundwork served as a site for an innovation in sustainable luxury, the remote underground passages have revealed an unusual Australian lifestyle.

Following, Anne Poelina and Johan Nordensvard's work "Sustainable Luxury Tourism, Indigenous Communities and Governance": the overall aim of this chapter is to explore the important intersection between traditional Aboriginal cultural and environmental management, knowledge and heritage, with the interest of sustainable luxury tourism in remote wilderness communities in Australia. The author argues that such a development could bridge the divide between culture and nature explaining how and why management and protection of landscapes and ecosystems are integral to human heritage, culture, and a new wave of sustainable luxury tourism.

Moving on, Feray Adıgüzel, Matteo De Angelis, Cesare Amatulli developed "Design Similarity as a Tool for Sustainable New Luxury Product Adoption: The Role of Luxury Brand Knowledge and Product Ephemerality," and they argue that luxury and sustainability are not incompatible concepts when luxury brands do employ the right product design strategy. Indeed, the effectiveness of two new green luxury product design strategies has been investigated in depth in this chapter. First, the green new product might be similar in design to luxury company's previous nongreen products. Second, the green new product might be similar in design to models of a different, non-luxury company specialized in green production. We investigate the effect of design strategy on new product purchase intention and propose that such an effect might be affected by fit (i.e., moderated mediated) by the combination of one product-related factor, such as product ephemerality (i.e., how long lasting the new product is), and one consumer-related factor, such as consumers' luxury brand knowledge (i.e., how much consumers know about the brand and its past).

Completing the book, Thomaï Serdari prepared a paper titled "The Carloway Mill Harris Tweed: Tradition-Based Innovation for a Sustainable Future." The Harris Tweed Industry, for most considered a remnant of the past, presents an opportunity to study its elements anew and reframe them both within the context of luxury and that of sustainability.

At the Carloway Mill, one of the three remaining "homes" where Harris Tweed is handcrafted, the authors studied the strengths, weaknesses, threats, and opportunities that have pushed a small enterprise to experiment with tradition in order to ensure a sustainable future for itself and its people. Specifically, they tried to understand how the Harris Tweed Industry arrived to where it is today and what it means for the Carloway Mill to have introduced a "new" type of fabric that addresses the modern wearer's concerns about climate change, environmental pollution, and zero-waste fashion design.

It is important to highlight that all of these diverse contributions represent a great step forward in expanding the insights in the field of sustainable management of luxury. It is certainly the most comprehensive collection of writings on this subjects to date. Note that this initiative has received a large international response, ant it is expected to continue to stimulate further debate.

Bibliography

Dixon SEA, Clifford A (2007) Ecopreneurship: A New Approach to Managing the Triple Bottom Line. J Organ Change Manag 20(3):326–345

Egri CP, Herman S (2000) Leadership in North American environmental sector: values, leadership styles and contexts of environmental leaders and their organizations. Acad Manag J 43.4:523–553

Frank RH (1999) Luxury fever—weighing the cost of excess. Princeton University Press, New Jersey

Gardetti MA (2011) Sustainable luxury in Latin America. In: Conference delivered at the Seminar Sustainable Luxury & Design within the framework of IE—Instituto de Empresa—Business School MBA, Madrid, Spain

Hart SL (1995) A natural-resource-based view of the firm. Acad Manag Rev 20(4):986–1014

Hart SL (2005) Capitalism at the crossroads. Wharton School Publishing, Upper Slade River

Hart SL (2007) Capitalism at the crossroads—capitalism at the crossroads—aligning business, earth, and humanity, 2nd edn. Wharton School Publishing, Upper Slade River

Hart SL, Milstein M (1999) Global sustainability and the creative destruction of industries. MIT Sloan Manag Rev 41(1):23–33

Hart SL, Milstein M. (2003) Creating sustainable value. Acad Manag Executive 17(2): 56–67

Rainey DL (2004) Sustainable development and enterprise management: creating value through business integration, innovation and leadership. In: Article presented at Oxford University at its colloquium on Regulating Sustainable Development: Adapting to Globalization in the 21st Century—Aug 8 through 13 2004

Ricca M, Robins R (2012) Meta-luxury—brands and the culture of excellence. Palgrave Macmillan, New York

World Commission on Environment and Development-WCED (1987) Our common future. Oxford University Press, Oxford

Contents

The Face of Culturally Sustainable Luxury: Some Emerging Traits from a Case Study

Silvia Ranfagni and Simone Guercini

Abstract This chapter investigates cultural sustainability in luxury. Seen as a path followed by luxury brands to increase the perception of sustainable luxury, the authors have identified its drivers, prerequisites and managerial implications. In doing so, they have studied the integration of cultural heritage and luxury through the representative case of a luxury company which produces fabrics expressing local traditions and uses them in its collections. The study demonstrates that culture combines in luxury in terms of symbols, knowledge sets, natural resources, behaviours and community spirit. These can be considered potentially determining factors in the fostering of uniqueness in sustainable luxury.

Keywords Sustainability · Luxury · Cultural heritage

1 Luxury and Sustainability: As Different as They Are Similar

The concept of sustainability has been examined in depth and widely discussed in the management literature, and its confines are continually expanding. Indeed, the concept is complex and multifaceted. The dimensions contributing to its composition have been identified as cultural, social, organizational and economic. Thus, sustainability in companies expresses itself in actions aimed at preserving natural

Silvia Ranfagni and Simone Guercini share the final responsibility for this chapter. However, Silvia Ranfagni wrote the Sects. 1, 2, 4 and 5; Simone Guercini wrote the Sect. 3. Silvia Ranfagni and Simone Guercini composed together the Sect. 6.

S. Ranfagni (✉) · S. Guercini
Department of Economics and Business, University of Florence,
Via Delle Pandette, 9, 50127 Florence, Italy
e-mail: silvia.ranfagni@unifi.it

S. Guercini
e-mail: simone.guercini@unifi.it

© Springer Nature Singapore Pte Ltd. 2018
M. A. Gardetti and S. S. Muthu (eds.), *Sustainable Luxury, Entrepreneurship, and Innovation*, Environmental Footprints and Eco-design of Products and Processes, https://doi.org/10.1007/978-981-10-6716-7_1

resources (Pullman et al. 2009), reducing waste in productive processes (Gordon 2007), creating more egalitarian societies in which poverty and disparity are reduced (Montiel 2008), and developing a green-based company ideology (Closs et al. 2011). The final objective of the company is, then, to transform these actions into positive sources of performance. However Seidman (2007) points out that in general "sustainability is about much more than our relationship with the environment; is about our relationship with ourselves, our communities and our institutions" (p. 58). It is, then, a social but also a personal issue. It lies in and implies a strong individual conviction that directly involves the people who work in a company. Moreover, the effects of a lack of sustainable behavior can be such that in today's companies sustainability is experienced collectively as an ethical question (Aguilera et al. 2007). It is a real but also a shared goal of business ethics (Crane and Matten 2016) as well as an enduring feature of a company's social responsibility (Joy et al. 2012).

It may seem paradoxical, therefore, to join sustainability with luxury. Sustainability in fact, is a *collective phenomenon* since preserving the environment means protecting a common good. Understood as "meeting the needs of the present without compromising the ability of future generations to meet their own needs" (Brundtland Commission Report 1987), it expresses an inevitable social need and, as such, incorporates the ethical values of altruism, reciprocal respect and moderation. On the contrary, according to Berry (1994) luxury is an expression of *social* and *moral transgression*. Its values are those of hedonism, expense, inessentiality and affluence. Thus, luxury is also *inequality*. In this regard, Douglas and Isherwood (1996) and Heine (2012) consider luxury consumption as a highly communicative act, signaling status, wealth, and class, thus an expression of sought-after social and economic power. However, luxury is also something else. De Barnier et al. (2012) highlight how luxury is not merely hedonism, exclusive character (prestige), high prices (expensive), but it is also *rarity*. Kapferer (2010) as well underlines this relation between rarity and luxury seeing in it an incontestable connection with sustainability. This too, then, is rarity. Luxury is based on the employment of rare resources (raw materials) and competences (know-how). Sustainability shields natural resources, which are physiologically scarce. What is more, just as luxury depends on sustainability in that it employs natural resources (albeit rare), in the same way it feeds sustainability. Indeed, it contributes to reducing the utilization of natural resources; this is because the high prices of luxury goods limit their demand on the part of consumers. In addition to rarity, as Kapferer (2010) shows, luxury and sustainability are characterized by *durability*. He claims that "ninety per cent of all Porsches produced are still being driven. Louis Vuitton provides after sales services to any genuine LVuitton product, whenever it was bought" (p. 42). Luxury lasts in time. In this sense it is similar to sustainability, which pursues the environment duration through its conservation. Instead of a contraposition between sustainability and luxury, a synergetic relation between the two emerges. This relation, however, cannot be taken for granted. Indeed, it is necessary to look at the other side of the coin, that is, the consumers and how they view the relation between luxury and sustainability. As Gardetti and Torres (2014) demonstrate, luxury consumers seem not to include sustainability as a criterion in determining whether to purchase luxury brands. More specifically, Davies et al. (2012) stress that consumers consider sustainability much less in their luxury consumption decision-making process than in their commodity consumption decisions, although luxuries are perceived as more sustainable if compared with commodity products. According to Ehrich and Irwin (2005) this scarce reactivity is due to the fear that sustainability could break the individual spell that luxury creates. Dekhili and Achabou (2013) show that this spell can become compromised if one of the links that holds luxury and sustainability together is broken, that is rarity. According to these authors, in fact, the customers may not value positively the idea of buying Hermès products made from recycled cotton. This is because recycling means that luxury is not so rare and thus loses its prestige. It is necessary, then, to create a connection between luxury and sustainability based on perceived rarity. Even in the event that such a link exists, it may not be sufficient to render sustainability a source of differentiation in luxury. Recent studies (Kapferer and Michaut-Denizeau 2014) point out that if, on the one hand, consumers react promptly when they receive news of luxury companies which adopt behaviors in contrast with environment protection, on the other, they remain indifferent to news that reveals the active participation of a luxury brand in sustainable activities. Consumers assume that luxury brands adopt sustainable behaviors. Such behaviors are considered part of their business mission. It is necessary, then, to render sustainability not as "an implicit need without having previously been an expressed one" (Berger et al. 1993). To this end it may be necessary to rethink the rarity link.

2 On the Trail of Diversity: From Brand to Cultural Heritage

According to some scholars, the link of rarity needs to be reformulated by acting on the luxury goods "to increasingly highlight product quality, their being handmade, their rare craftsmanship" (Kapferer and Michaut 2015, p.15). The consequence would be a rarity differentiating sustainable luxury based on the uniqueness of *competences*, *abilities* and *know-how* that are involved in its realization (Fionda and Moore 2009). Luxury would contribute to preserving also rare immaterial resources, which are indispensable for its continuity in time. And sustainability would assume intangible qualities of a cognitive nature. It is to these qualities that luxury businesses could consider to strengthen the concept of quality offered pursuing an aim which Kapferer and Michaut (2015, p. 15) express in these terms: "Luxury is by definition the highest quality; [now] it has to redefine quality". Nonetheless, Kozinets and Handelman (2004) assert that "while the dream quality is essential to a luxury product, in some instances, a long history and heritage further intensify a [luxury] brand's strength". Thus, a luxury brand can use its *heritage* as a basis for its diversity. Since heritage is a carrier of historical values from the past (Nuryanti 1996), the uniqueness it creates lies in the historical values of the brand. This uniqueness can tie luxury to sustainability and become driver of the differentiation of sustainable luxury. In order to do this, it has to be genuine and not merely evoked. More specifically, it has to lie not only in the narratives about the origins and evolution of the brand (Aaker 2004), but also in the continual demonstrations of historical significance for the brand. Such demonstrations consist in showing how the original brand values have not been lost but constitute still today the distinguishing foundation of the brand (Peñaloza 2000). There follows the surfacing of a *continuum* of brand values that mark the journey of self-authentication undertaken over time by the brand (Postrel 2003). In this resulting *authenticity* lies the connection between luxury and sustainability. In other terms, luxury intertwines with sustainability by way of authenticity. Pursuing an identity over time without renouncing the nature of one's origins means preserving what is authentic, which today constitutes a prestigious social value (Firat and Venkatesh 1995; Ranfagni and Guercini 2014) and which, as such, assumes the quality of a collective good. Luxury based on brand heritage can contribute to the conservation of what is authentic and makes of it a bedrock of its rarity (Urde and Balmer 2007).

However, it may also occur that in addition to preserving a "business past" a company can also contribute to conserving a "social past". In other words, certain recurring brand values may become a means of safeguarding a cultural heritage. Defined as a composite of history, coherence and continuity of a group's or society's distinguishing characteristics (Arantes 2007), the cultural heritage describes ways of living developed by a community and passed on from generation to generation, including customs, practices, places, objects and artistic expressions. It also consists of values and traditions. As some studies have demonstrated (Hakala et al. 2011), when it is embedded in a brand heritage, the brand can preserve in its authenticity certain collective values. The brand is integrated in these values, and they enhance its identity. This combination gives rise to what we could call *cultural sustainability*. This presupposes a process of self-authentication that the brand or the product pursues over time and which intertwines with a cultural heritage. Cultural sustainability is not a momentary phenomenon but must be built up over an arc of time. It implies a thorough search for identity. This issue has not yet been explored in the literature on luxury. Luxury is studied in its ties with the culture of the country of origin in terms of cultural stereotypes (e.g. French luxury vs Italian luxury) (Corbellini and Saviolo 2014). More intrinsic relations between history, culture and tradition have been identified in more recent studies on ethnic apparel. This clothing is seen as "ensembles and modifications of the body that capture the past of the members of the group, the items of tradition that are worn and displayed to signify cultural heritage" (Chattaraman and Lennon 2008, p. 521). The recovery of the past takes place through the product; it incorporates, in fact, an ethnic culture. Similar cases emerge from the histories of luxury companies. Some of these (Gucci, Zegna, Prada, Ferragamo, Tod's) in fact seek to differentiate themselves by creating collections that employ raw materials which are expression of a local culture; in doing so they also contribute to its conservation. The incorporation of a local culture usually involves resorting to semifinished textile products or clothing traditions (ethnic apparel) that are intertwined with that culture. Here the perceived uniqueness lies in the *integration* one manages to generate between the product and the cultural heritage that it represents. Integration underpins the highest grade of cultural sustainability. The authentic that derives from it is a common good that is born of values that are universally recognized as characteristic of a collective culture. The ensuing effect is to "generate transmission from generations to generations of timeless products" (Kapferer and Michaut 2015) as expression of an integration with a historical heritage. The more the product is integrated in the heritage, the more it preserves the heritage and the more it can be perceived as rare. In virtue of this, it becomes essential to understand how culture can enter luxury. In other words, it is necessary to determine (a) the drivers of cultural sustainability in luxury (in what are culture and luxury integrated), (b) how a local culture can be embedded in luxury, and (c) the managerial implications for brands that intend to pursue cultural sustainability. To this end we will investigate a model case in which luxury is rooted in cultural heritage. We will begin with the history of a cloth that was created as expression of a local culture and go on to examine how the local company that produces that cloth today has managed to integrate a local cultural heritage in its luxury collections.

3 Research Methodology

Our research comprised two phases of analysis. First we gathered the secondary data (Eisenhardt and Graebner 2007) necessary to identify productions of clothing as expressions of a local culture. In this phase we identified three territorial areas of interest within the setting of the Italian textile and clothing market (Casentino in Tuscany, Val Pusteria in Trentino and Sassari in Sardinia). Each of these areas is distinguished by the presence of producers that manufacture typical fabrics expressing local history and traditions and use them to create their collections. In our research we focused on the area of Casentino and on the fabric that bears its name. This, if compared with fabrics of other areas, distinguishes itself as highly original, quite rare and deeply rooted in the territory. Then, we applied case study methodology (Yin 1989) to investigate one company, Tessilnova, which can be considered a representative case for our study objectives. In fact, it has been able to preserve strong ties with Casentino fabric, local culture and its collections. The case was reconstructed on the basis of research activities that have engaged us for over three years. Like anthropologists we have studied the historical documents describing the company and its business and we have participated in its locally organized promotional events. We have also spent time observing its employees during the production of the fabric and of the collections. All the information gathered was integrated with that gained from interviews carried out with the owner and his collaborators. The issues examined were: (a) the history of the company; (b) the relation between the company, its production and the local culture; (c) the process of authenticating in the collections the culture embedded in the local fabric employed; (d) the relative impacts on production, organization and marketing; and (e) the paths of growth taken by the company with the aim of preserving the link between culture and collections. The interviews were transcribed, explored and interpreted by the authors (Krippendorff 2004) and integrated with the secondary data. However, before proceeding with the analysis of the case, we provide information about the historical nature of the Casentino fabric and the correlated local culture.

4 A Cultural Heritage Preserved in Tessilnova Collections

4.1 Casentino Fabric: The Interweaving of Local and Textile Culture

Casentino fabric is produced in Tuscany (Italy), more precisely in the mountainous region known as Casentino at the beginning of the Tuscan-Romagnolo Apennines. The region includes 12 towns, stretches over an area of 826.49 km^2, and is recognized as a territory with marked cultural and naturalistic potential and as an attractive destination for tourists. The small towns of Camaldoli and La Verna are known as centers of Catholic religiosity. Also well known are the archaeological treasures, composed of Etruscan sites, Romanesque churches and medieval manors. To these attractions must be added the nature trails interlacing the National Park of the Casentino Forests. In the past Casentino was a land of pastures used for flocks of, above all, indigenous sheep. From the XV century their wool was a source of real wealth for the area; driven by the flourishing textile center in Florence, it was marketed in many countries around Europe. In addition to the wool, another important local resource consisted in the waterways. Employed as the diving force to run textile machinery, they contributed to providing a powerful stimulus to the Casentino wool industry. The industry reached its peak in the XIX century when it gave employment to hundreds of local inhabitants. The product which characterized this industry was precisely Casentino fabric. Although there are historical documents that mention the cloth as early as 1272, its creation is not formally attested until 1537. Referred to as a fabric "for princes and carters", it was very resistant, impermeable and well adapted for those who spent long periods of time outdoors. The fabric was ideal for cloaks worn by shepherds, merchants and local carters who needed garments that would hold up under bad weather and prolonged use. Its distinguishing feature is the curl, clearly visible on the surface of the cloth. This is obtained through a specific finishing procedure, teasing, which follows the extraction of the wool, using the technique of napping. The competences of the local wool industry developed specializations around the creation of the Casentino fabric, which in the wake of innovations in production (new machinery, aesthetic improvements) has undergone constant change over time. From a rustic, rough and coarse cloth it turned into a fabric of pure wool, light, warm and elegant. The brown and grey of its origins were joined by goose beak orange, which remains the distinctive color of the fabric. It became widely used by nobles and prominent personalities and was especially employed in the sewing of cloaks and double-breasted overcoats, with martingale belt and fox collar, a symbol of elegance and refinement. Today it is an accessible luxury, as is witnessed by its widespread use among diverse social classes. Recently, however, the local wool industry which produces the fabric has changed face. The flight from the countryside of the last century, the urbanization of the large cities, and not least the competition from countries with low cost labor, all these factors have led to a reshaping of the industry. There is only a minimal presence of native sheep in the area and fewer phases of the production of the fabric are carried out locally. There are two local businesses that produce and market the cloth, Tessilnova and Tacs. It is the intention of both to preserve a cloth steeped in traditions. Today it is perceived as authentic and as a niche product. In virtue of these qualities it attracts many fashion designers (Gucci, Stefano Ricci and others), who employ it in their collections.

4.2 Tessilnova: The Case of a Luxury Embedding History and Local Traditions

The business was established in 1961 when Gabriele Grisolini purchased two looms from the Antico Lanificio, a company founded in 1864, where he served an apprenticeship. Grisolini continued to produce Casentino fabric for this company until the seventies when, on the arrival of new owners (the Morelli family taking the place of the Lombards), he decided to search for new clients. In 1975 he obtained a license for the direct sale of fabrics and finished articles of clothing, consisting in particular of shawls and scarves. In 1982 he opened a factory shop. And subsequently he purchased a significant portion of the premises of the Antico Lanificio (which had in the meantime ceased activity). These spaces were used to increase the areas dedicated to both production and sales. Developing a network of collaboration with local tailor's shops, the company began to make gilets, overcoats and cloaks. Today it is still a family business, boasts five operators, and offers collections of overwear (jackets, overcoats and gilets) for men, women and children in addition to a wide range of accessories (bags, shoes, scarves, etc.). The precious collections are sold in their own shop situated in Stia (Casentino) as well as in luxury multibrand stores in Italy and abroad. All the articles are made from Casentino cloth produced by the company. The quantities of its production oscillate between 150 and 200 bolts a year, each about 50 m long. The colors used amount to at least thirty. The cloth is reserved not only for the company's own collections but also for important luxury designers intent on meeting the requirements of customers who seek in luxury not so much the Made in Italy as local history and traditions. Remarkable efforts in this direction are being made by Claudio Grisolini, current owner and son of Gabriele, who strives to reinforce in Tessilnova collections the tie with the territory.

This is readily recognizable by the *curl* emerging from the fabric. Its production requires the employment of specific manufacturing techniques. As a consequence of the dismemberment of the local wool industry, not all these techniques are available locally. Tessilnova is thus forced to search for suppliers located outside of Casentino, selecting them on the basis of their manufacturing specializations. The preference is given to those provided with the competences necessary for creating a cloth that technically reproduces Casentino fabric. These skills are found in the nearby district of Prato, which is home to highly qualified textile firms. Thus, the phases of spinning and carbonization are performed by some of these district companies. As to the choices, the owner explains that: "The important thing is to preserve the fabric as a product of a traditional manufacturing process ... the search for specialized competences serves this end...." The weaving, on the other had, is done by Tessilnova. The dying and finishing operations, although not carried out internally, are locally circumscribed. They are realized, in fact, by businesses located in Casentino. Manufacturing specialization is combined with continuous researches into the *manufacturing origins* of each single fabric that Tessilnova creates. It results an original and an authentic luxury. It is just this luxury that distinguishes Tessilnova from its main competitor. The quest for originality is carried out by exploring codified knowledge, which is handed down over time. In this regard, the owner enjoys pointing out what follows: "I've the good fortune to revisit certain aspects of the manufacturing production thanks to an old pattern book I inherited from my father. This contains in addition to the swatches, the technical descriptions (number of threads, frame, etc.) relative to each single fabric that the Antico Lanificio produced up until the Second World War". Making a fabric drawing inspiration from this knowledge bank means creating a product which, albeit tacitly, embodies a local cognitive heritage that remains alive despite the passage of time. Its realization has required investments in the manufacturing process. The thread count that Tessilnova uses is in fact quite fine; in other words, the number of threads (both weft and warp) involved is high (about 500 threads more for each bolt compared with those employed by the competitor). This has made it possible to vary only slightly the original height of the fabric (having increased from 1 m 40 cm to 1 m 50 cm) and to avoid using synthetic fibers. In fact, the modern equipment along with the employment of specific oils are able to harmonize the fiber of the wool without mixing it with synthetic products. As a consequence, creating authentic luxury has implied manufacturing choices oriented toward achieving quality to the detriment of economy. *Quality* is a feature that Tessilnova seeks in *raw materials* as well. Specifically, as regards materials the company follows two paths. In the one hand, it seeks to recover and enhance the value of wools coming from the rearing of native sheep still present in the territory. On the other, since these wools are currently scarce, it searches for them elsewhere. Indeed, the company is carrying out an intense activity of scouting in Australia and in South America. It selects the materials using as yardstick the fineness and the length of the indigenous wools. These wools are fine with a fiber in part long and in part short. The search for reproductions of authentic wools outside Casentino becomes a way of maintaining a connection between natural resources and the territory. This connection lies in the commodity-based homogeneity between the wools produced locally and those found elsewhere. Promoting the recovery of native wool is, in any case, a goal that Tessilnova intends to pursue in the future. Its owner relates that "recently I managed to find over 300 kg of wool from indigenous wools to make indigenous garments ... I would like to help promote the return of Casentino sheep to our territory". The relation he establishes with these wools is almost sacral. "I guard the bolts we produce like an oracle. The price of the garment that comes out is usually much higher".

Overall, Tessilnova is a company that respects the *tradition* without distancing itself from *modernity*. That is, its collections are modern and also traditional. This combination is the result of the desire to renew of the owner. In addition to managing the business, he likes to experiment with conceiving new collections. Thus, he is a manager, but also a designer. In doing this he recovers the past in products which are at the same time modern. He does not work alone, but with designers who, following his orientations, develop models and prototypes of finished garments. These models are conceived from a historical perspective. The double-breasted coat is considered the most historical garment and the one that has inspired the creation of many articles of outerwear that Tessilnova offers the market today. The openness toward what is new has led the company to differentiate not only its models but also the colors of its collections. Beginning with goose beak orange, it has developed a color card with many alternatives. In this case, too, the owner acts as creator. The new colors, in fact, are fruit of personal experiences not so much with the company's manufacturing past as with the natural environment. In devising them, the owner drew inspiration from the hues of the woods of Casentino that he attends during his free time. What is more, using to advantage the warmth and the impermeability of Casentino fabric, a wide range of accessories such as scarves, bags, hats and gloves have enriched Tessilnova collections. But it does not end here. The company has also produced technical clothing designed to be used in particular situations. In this connection the owner states: "Our cloth enabled us to create a ski suit to be worn in snowy environments. This suit was developed together with researchers at the National Research Center and the University of Florence. Technical comfort analysis were carried out even on Monte Rosa to test the clothing at low temperatures".

Despite this bond between tradition, modernity and innovation, the clothes composing Tessilnova collections remain the clothes of a *local community* that, regardless of social changes, is made up of people who continue to identify in them. The sense of belonging that ties the members of the community to one another is intrinsic to Casentino fabric. It is revealed when, on visiting the company, you interact with the collaborators. It also transpires, as the owner points out, from the Casentino Museum of wool, which seems to sanctify the local culture and the traditional wool industry to which it is bound. The sense of community behind Casentino fabric contaminates to some extent the business's customers as well. Among them, who most deeply experience it are not the customers who come to know the fabric through the advertisement the company makes in specialized magazines or through multi-brand boutiques where it sells its collections, but rather the customers who get to know the territory of Casentino as tourists. These people, like the members of the local community, live that territory, even if only temporarily, experiencing it actively and filtering it through their own personal experiences. A number of experiential itineraries are offered with the aim of creating this sense of participation. The activity of Tessilnova is particularly intense in this connection. "We organize experiential itineraries to help people get to know the territory and its traditions. They're organized here in the company, which becomes a sort of mirror of the historical roots of the cloth, and in the nearby Museum, where you can really get to understand the history and the production techniques of Casentino fabric". Today Tessilnova stands out as a producer of luxury, which is culturally rooted in the territory. In doing this, it has to confront with new (not local) competitors that attracted by the possibility of "selling" the heritage, reproduce Casentino fabric although they lack adequate competences, thus risking its falsification. Preserving authenticity is the main challenge that a family business like Tessilnova must face. This challenge is certainly not easy given that the emerging players are hungry for new business and can count on huge financial and marketing resources.

5 Discussion

Our research makes it possible to identify which factors can constitute culturally sustainable luxury (in what way culture emerges in luxury), what conditions are necessary such that these factors take shape (how luxury manages to preserve culture) and the correlated choices businesses find they have to make (what it means to be culturally sustainable). Now, we discuss the main results that are synthetized in Table 1.

Table 1 Culturally sustainable luxury: components, conditions and implications

Constitutive factors (in what)	Prerequisites (how)	Business choices (What it implies)
Symbols as tangible elements evoking a local culture	Specialization in manufacturing	Outsourcing as production choice justified by the search for specialized competences
Local knowledge sets	Access to individually mediated codified knowledge	Leaning toward quality to the detriment of economy
Natural resources	Access to native and non-native wools	Production of the authentic by using evocative reproductions of the authentic
Entrepreneurial inclination to experiment	Circular relation with product ("to take care of")	Recover tradition in modernity
Community spirit	Inhabiting	Involving consumers in experiences of consumption locally organized

From the case study it emerges that a local culture penetrates luxury through its more tangible identifying features, that is, its *symbols*. They constitute the most visible part of the culturally sustainable luxury. It is they that create an immediate connection between products, territory and culture. The curl, in the case under examination, is what creates a natural tie between Casentino collections and certain local traditions. Nonetheless, it cannot be taken for granted. Contextual factors reduce the availability of local manufacturing competences and may drive to evaluate options of outsourcing phases of production. In view of this situation one could ask whether preserving certain traditional features by involving external firms to realize parts of a manufacturing process that had always been locally centralized, could lead to a reduction in the cultural value of the resulting products. In the need for specialization there is the solution to the emerging trade-off between in sourcing and outsourcing in production. In other words, the cultural value of luxury is preserved if it is fruit of specializations which, while external, are capable of reproducing the technical competences that are no longer available locally. Specialization then becomes a factor influencing the cultural sustainability: in the continuity of certain technical competences lies the continuity of a culture that can be transmitted by way of a "local luxury".

But there is more. The local culture remains alive in the *knowledge bank* that is incorporated in luxury. This knowledge bank is preserved by means of luxury. It is the most profound component of a culturally sustainable luxury, and therefore the least visible to the eyes of luxury purchasers. Moreover, it emerges from lasting codified manufacturing skills and know-how. These skills and knowledge constitute that tacit cognitive heritage which can foster the eternity of luxury once it becomes a part of it. The ability to integrate them in luxury, in fact, is the real challenge that businesses have to face. To do so it is not sufficient to decode and interpret them. In fact, they do not remain immobile as remnants of a past history, and become transferable if mediated by a creative and productive human sensitivity. This sensitivity is a feature of the passionate entrepreneur who, having a role in the generation of collections, creates an authentic luxury that is individually filtered. The creation of a luxury that elegantly incorporates deeply rooted knowledge is in line with entrepreneurial behaviours in which the leaning toward quality wins out over the search for economy. Ultimately, the potential for luxury to be culturally sustainable depends on its capacity to incorporate certain knowledge sets and to foster them over time.

Another important factor in culturally sustainable luxury consists in the *natural resources*. In particular the local culture is transmitted in luxury by way of the natural resources employed (the wools), which give expression to the culture. These resources may or may not be native. In other words, they may be produced locally and thereby be a direct expression of the local culture. On the other hand, they can also be sought beyond the confines of the territory. Of the latter kind, resources will be selected which share significant characteristics with the indigenous ones. These similarities will render them evocative of the local culture. Since authenticity is found in the "original and real things", but also in "authentic reproductions" (Peterson 2005, p. 208) and thus in "something whose physical manifestation resembles something that is original" (Grayson and Martinec 2004, p. 298), the use of non-native wools does not decrease the authentic value of the resulting product, nor does it negatively impact its cultural value. Evoking a territorial culture, even by means of wools reproducing the original ones, is a sign of the desire to preserve in the product a profound naturalistic identity which embodies the relation between the territory and the natural resources employed (wools). It follows that cultural sustainability can also be environmental sustainability; that is, the conservation of natural resources depends also on the conservation of a local culture.

Culturally sustainable luxury, however, also includes the *entrepreneurial inclination* to experiment and renew. Luxury can preserve this attitude, which is intrinsic to local culture. This is because it is characteristic of those who contributed more than others to shaping that culture: the peasants. Peasants establish a circular relation with their productions: they take care of them (as the doctor takes care of a patient) and they follow them as they evolve to achieve certain manufacturing aims while exploiting their innate sense of experimentation. The collections as well reflect the efforts of searching and experimenting. The historical values these collections incorporate do not annul these efforts, which require openness toward what is new in the respect of the past. Modernity, that is, is not antithetic to traditions but rather resides in them. This means that new articles of clothing are created by drawing inspiration from the historical ones and ultimately have evolved from them. Their creation may involve the owner who, embodying the local culture, tends more spontaneously to follow an inclination to experiment. He can be also animated by a sense of personal pride that comes from feeling himself ambassador of culture to which he is belong and which he tries to transmit through his business activity.

Finally, culture enters luxury through the *spirit of community*. This is another component of cultural sustainability that luxury can conserve. Whoever purchases Casentino collections, in fact, buys products that take shape within community contexts. These contexts, constituting part of the local culture, become mediator of its values. The spirit of community does not remain impalpable. The customers, in fact, gain awareness of it when they have the opportunity to experience the community. These customers are above all the tourists who, immerging themselves in locally organized experiences, can live these communities. They live them in the sense of the Latin "habitare" (Guercini and Ranfagni 2016), that is, spending time with locals and sharing collective spaces that transmit a sense of community belonging. Enabling the "inhabiting" of contexts consists in organizing experiential itineraries where the union between community and fabric emerges in all its clarity. These involve the company, with its local collaborators, and the museums, which narrate the history of wool and of the territory.

6 Conclusions

The case we have examined is one of luxury, which has maintained its authenticity over time (in that it incorporates its heritage), but at the same time it has turned traditions and local values into its distinguishing traits (it also incorporates a cultural heritage). It is an example of a luxury at once authentic and culturally sustainable. Its uniqueness consists in its integration with a local culture. The prerequisite for this integration is the existence of a textile or clothing artefact whose history is intertwined with that of a collective culture. The creation of a culturally sustainable luxury can stem from such a condition. It is not surprising that businesses, like Gucci, who have begun to experiment with culturally sustainable luxury, have decided to create collections using culturally rooted fabrics (like loden and Casentino fabric). They have sought, that is, to initiate a process which combines their brand heritage with a cultural heritage and that would be well worth examining in depth. Before doing this, however, it is necessary to investigate how a culture is integrated with a luxury, beginning with the analysis of cases (like ours) in which the intertwining is pure and innate, so that one can then observe how the resulting levels of integration relate with the history and values of luxury brands. This relation determines the cultural value of the brand and thus its level of cultural sustainability.

Our study shows that culture enters luxury in terms of symbols (the curl), knowledge sets, natural resources (even if only evoked), ways of being and community spirit. The constitution of culturally sustainable luxury, then, revolves around the dimensions of manufacturing, cognition and attitude, and identity. They are different from those of conventional sustainable luxury. The *manufacturing dimension* includes the symbols and the natural resources. The cultural value of a luxury product is dependent upon both. The reproductions of cultural symbols (the curl) require certain competences and manufacturing techniques that may be sought outside the original territory. Provided they are specialized, in fact, they do not detract from the cultural value of the product. This also happens when the natural resources employed (native wools) are non-native and constitute an evocative reproduction of the original. In fact, it is assumed that the consumer perceives as authentic also what is merely evoked as authentic. The *cognitive dimension* is submerged. It includes knowledge sets and ways of being, which means that in these, too, the cultural value of luxury can lie. Luxury, then, depends upon access to deep-seated, culturally coded sets of knowledge, which make their way into the product. What is more, it relies on the ability to create a product stemming from an individual characteristic (inclination toward experimentation), which is in turn, expression of the local culture. This is a dimension that the purchaser can perceive only in the effects it produces, that is in the constant leaning toward the quality of the product and an ability to combine tradition and modernity in the collections. The first of these effects translates into a greater use of raw materials. The *community dimension* is composed of the community spirit. The cultural value of luxury resides in the community identity it is able to incorporate. It is not easy to transfer this identity. The transfer requires making the customer part of the community. He must experience it seeking to immerge himself in it. Only in this way is it possible to make him perceive his tie with the product and create a mental association with it. If we compare the three dimensions there emerges a situation that could appear to be contradictory. In particular, it seems that if on the on hand culturally sustainable luxury contributes to preserving natural resources, on the other the high quality that distinguishes it makes a greater use of these resources indispensable. But to the extent to which higher quality guarantees a greater durability of the product, the environmental sustainability intrinsic to culturally sustainable luxury can remain unaltered. There is another important observation to make. When we raise the components of the three dimensions to a higher level of abstraction, we find more *universal values intrinsic* to culturally sustainable luxury. These values are realism (reproduction of symbols in conformity with what is real), environmentalism (protection of natural resources), immaterialism (the recovery of knowledge sets and ways of being), and communitarianism (sense of community). A luxury brand engaged in projects aimed at recovering a local culture intertwined with a manufacturing culture (textile and clothing) will need to explore the integration of these values and its own values. This *valued-based integration* is preliminary to the more operative one, which involves the integration of the components of the three dimensions of culturally sustainable luxury. It is on the foundations of this second integration that a luxury brand will then build its perceived rarity over time. However, it is quite likely that the three dimensions coexist, especially they rest on deeply rooted, local entrepreneurship, made up of people who express the culture which is embodied in the luxury. It is this situation that renders culturally sustainable luxury brands more authentic and, as such, more defensible. In any case, the analysis of the relation between luxury brand heritage and cultural heritage will be one of the aims of our future research. One of the limitations of the present study consists in the limited extension of the empirical analysis and in its exploratory nature. Indeed, we focused on only one, albeit representative, case of culturally sustainable luxury. We intend to expand our research to include other cases so as to identify other dimensions that constitute culturally sustainable luxury and thus to discover their founding abstract values. The ultimate objective is that of determining spheres in which luxury and culture establish a dialectical and lasting relation, all this in a relation of reciprocity: luxury can in fact thrive in culture, but also the opposite may occur. The real challenge is to try to understand how. In facing this challenge managers, anthropologists, historians and artists can all make a contribution. The issue, in other words, can clearly take on an interdisciplinary character.

References

Aaker DA (2004) Leveraging the corporate brand. Calif Manag Rev 46(3):6–18

Aguilera RV, Rupp DE, Williams CA, Ganapathi J (2007) Putting the S back in corporate social responsibility: a multilevel theory of social change in organizations. Acad Manag Rev 32(3):836–863

Arantes AA (2007) Diversity, heritage and cultural politics. Theor Cult Soc 24(7–8):290–296

Berger C, Blauth R, Boger D, Bolster C, Burchill G, DuMouchel W, ... and Timko M (1993) Kano's methods for understanding customer-defined quality. Center Qual Manage J 2(4):3–35

Berry CJ (1994) The idea of luxury: a conceptual and historical investigation, vol 30. Cambridge University Press

Chattaraman V, Lennon SJ (2008) Ethnic identity, consumption of cultural apparel, and self-perceptions of ethnic consumers. J Fashion Mark Manage: An Int J 12(4):518–531

Closs DJ, Speier C, Meacham N (2011) Sustainability to support end-to-end value chains: the role of supply chain management. J Acad Mark Sci 39(1):101–116

Corbellini E, Saviolo S (2014) Managing fashion and luxury companies. Etas

Crane A, Matten D (2016) Business ethics: managing corporate citizenship and sustainability in the age of globalization. Oxford University Press

Davies IA, Lee Z, Ahonkhai I (2012) Do consumers care about ethical-luxury? J Bus Ethics 106(1):37–51

De Barnier V, Falcy S, Valette-Florence P (2012) Do consumers perceive three levels of luxury? a comparison of accessible, intermediate and inaccessible luxury brands. J Brand Manage 19(7):623–636

Dekhili S, Achabou MA (2013) Price fairness in the case of green products: enterprises' policies and consumers' perceptions. Bus Strategy Environ 22(8):547–560

Douglas M, Isherwood B (1996) The world of goods. Routledge, New York

Ehrich KR, Irwin JR (2005) Willful ignorance in the request for product attribute information. J Mark Res 42(3):266–277

Eisenhardt KM, Graebner ME (2007) Theory building from cases: opportunities and challenges. Acad Manage J 50(1):25–32

Fionda AM, Moore CM (2009) The anatomy of the luxury fashion brand. J Brand Manage 16(5–6):347–363

Firat AF, Venkatesh A (1995) Liberatory postmodernism and the reenchantment of consumption. J Consum Res 22(3):239–267

Gardetti MA, Torres AL (eds) (2014) Sustainable luxury: managing social and environmental performance in Iconic brands. Greenleaf Publishing

Gordon DK (2007) Using lean to meet quality objectives. Qual Progress 40(4):55

Grayson K, Martinec R (2004) Consumer perceptions of iconicity and indexicality and their influence on assessments of authentic market offerings. J Consum Res 31(2):296–312

Guercini S, Ranfagni S (2016) Conviviality behavior in entrepreneurial communities and business networks. J Bus Res 69(2):770–776

Hakala U, Lätti S, Sandberg B (2011) Operationalising brand heritage and cultural heritage. J Prod Brand Manage 20(6):447–456

Heine K (2012) The concept of luxury brands. Luxury Brand Management, no. 1, 2edn, ISSN: 2193–1208. http://www.conceptofluxurybrands.com

Joy A, Sherry JF Jr, Venkatesh A, Wang J, Chan R (2012) Fast fashion, sustainability, and the ethical appeal of luxury brands. Fashion Theor 16(3):273–295

Kapferer JN (2010) All that glitters is not green: the challenge of sustainable luxury. Eur Bus Rev (November–December), 40–45

Kapferer JN, Michaut-Denizeau A (2014) Is luxury compatible with sustainability? luxury consumers' viewpoint. J Brand Manage 21(1):1–22

Kapferer JN, Michaut A (2015) Luxury and sustainability: a common future? The match depends on how consumers define luxury. Luxury Res J 1(1):3–17

Kozinets RV, Handelman JM (2004) Adversaries of consumption: consumer movements, activism, and ideology. J Consum Res 31(3):691–704

Krippendorff K (2004) Content analysis: an introduction to its methodology. Sage

Montiel I (2008) Corporate social responsibility and corporate sustainability separate pasts, common futures. Organ Environ 21(3):245–269

Nuryanti W (1996) Heritage and postmodern tourism. Ann Tourism Res 23(2):249–260

Peñaloza L (2000) The commodification of the American west: marketers' production of cultural meanings at the trade show. J Mark 64(4):82–109

Peterson RA (2005) In search of authenticity. J Manage Stud 42(5):1083–1098

Postrel V (2003) The substance of style: how the rise of aesthetic value is remaking culture, commerce, and consciousness

Pullman ME, Maloni MJ, Carter CR (2009) Food for thought: social versus environmental sustainability practices and performance outcomes. J Supply Chain Manage 45(4):38–54

Ranfagni S, Guercini S (2014) On the trail of supply side authenticity: paradoxes and compromises emerging from an action research. J Consum Behav 13(3):176–187

Seidman D (2007) How: why how we do anything means everything … in business (and in life). Wiley, New York

Urde M, Greyser SA, Balmer JM (2007) Corporate brands with a heritage. J Brand Manage 15(1):4–19

Yin RK (1989) Case study research: design and methods (Rev. ed). Newbury Parks

How the Business Model Could Increase the Competitiveness of a Luxury Company?

Elisa Giacosa

Abstract The luxury business is one of the most important industries in many countries. In the recent years, it is affected by some paradigms (such as consumption crisis, democratization process, market globalization) that should be considered while defining the company business model. To overcome these paradigms, the business model has to be characterized by a set of factors which could increase the competitiveness of a luxury company. The objective of this research is to verify how the business model could increase the competitiveness of a luxury company. We used a qualitative approach; in particular, we considered large and medium-sized luxury companies which are internationally-recognized with highly innovative and entrepreneurial business approach over generations. The research identifies a series of business model's characteristics for ensuring or increasing a long-term competitiveness, in terms of innovativeness, company's assets and long-term vision. Key factors identified above should be applied together to understand better the phenomenon of innovative luxury business.

Keywords Luxury sector · Luxury business · Innovation strategy
Innovativeness · Company's assets · Long-term vision

1 Introduction

The companies operating in the luxury sector can run their activity in two luxury levels: non-affordable and affordable luxury. This sector is characterized by several paradigms that impact on the company's business models (Giacosa 2016).

The first paradigm refers to the consumption crisis: the company has to develop new business model or improve the existing one, in order to increase its competiveness and attract the market's attention.

E. Giacosa (✉)
Department of Management, University of Turin,
Corso Unione Sovietica, 218 Bis, 10134 Turin, Italy
e-mail: elisa.giacosa@unito.it

© Springer Nature Singapore Pte Ltd. 2018
M. A. Gardetti and S. S. Muthu (eds.), *Sustainable Luxury, Entrepreneurship, and Innovation*, Environmental Footprints and Eco-design of Products and Processes, https://doi.org/10.1007/978-981-10-6716-7_2

17

The second one relates to democratization of the luxury (Silverstein and Fiske 2003). A lot of luxury companies—by reducing products exclusivity—have permitted a wide range of customers to take advantage of purchasing luxury products. In this context, the luxury gets into affordable luxury levels and, as a result, the luxury suffered on its integrity (Thomas 2007). Sometimes, it also deteriorates the reputation and legitimacy of the luxury companies (Giacosa 2012).

Finally, the market globalization affects the pureness of luxury by spreading production processes to another countries, which can decrease the quality of luxury products and it can have a negative impact on the brand image. An example constitutes the case of haute couture in France, which deterioration has started from outsourcing the production processes in foreign countries. That is why the luxury companies should be aware of the risk connected with deterioration of their luxury status. They should strengthen their origin country factor (Kapferer and Bastien 2009; Corbellini and Saviolo 2009).

To overcome these paradigms, the company's business model has to be characterized by a set of factors which could increase the competitiveness of a luxury company. Because of these reasons, numerous researchers focused on developing different business models which could improve competitiveness of luxury company. In this context, an innovative business model has been considered as a new paradigm, and it provides different way in reaching or maintaining the competitive advantage in the long term vision (Giacosa 2012). It should be a driving force in reaching better company's performance, environment and social development, cultural art and innovation of several nationalities, and conserving local craftsmanship (Gardetti 2011).

Regarding the literature, innovative business models in a large and medium-sized luxury sector have not been widely discussed. Present publication is therefore aimed at filling this gap and to verify if development and adaptation of innovative solutions can help the company to achieve a competitive advantage.

The objective of this research is to verify how the business model could increase the competitiveness of a luxury company. Its additional value is linked to the fact that the research identifies a series of business model's characteristics for ensuring or increasing a long-term competitiveness.

In analysis a qualitative approach was used. We analyzed different large and medium-sized luxury companies which are internationally-recognized with highly innovative business approach over generations.

The contribution is structured in the following way. The first paragraph represents description of the luxury business in division at core and no-core business, and division of luxury at affordable and non-affordable one. In the second paragraph, the issue of innovation approach in large and medium-sized companies is discussed. The third paragraph introduces the business model that the companies can use to improve its competitiveness and customers awareness about innovative products. Finally, the conclusions and implications of the study are set out, along with the limitations of the research.

2 Background

It is difficult to define the luxury context as the literature introduces many different definitions for the following reasons (Chevalier and Mazzalovo 2008; Corbellini and Saviolo 2009; Giacosa 2012, 2016; Jackson 2004):

- the luxury product is differently perceived by each individual;
- context and sphere of action impact the social value of luxury products;
- luxury products are spread among many sectors that can be simply divided into core luxury sectors (traditional ones like for example clothing, accessories, jewelry, perfumes and cosmetics, cars) and new luxury sectors (that appear in the luxury market: tourism and catering, wine, spirits and other gourmet products, furniture and household items). Such a division allows the customers to demonstrate their membership to a certain social class.

In order to present a complete contextual framework suited to the customers' requirements, the luxury business has been distinguished by brand affordability factor into two following categories (Chevalier and Mazzalovo 2008; Giacosa 2011, Giacosa et al. 2014; Okonkwo 2007):

- non-affordable luxury business: because of their high prices, products are not available for wide range of customers but just for a selected group of them. Those products are characterized by high quality and creative design, they are made in limited editions, often customized and tailored giving the customer a high social status. In the context of jewelry, the example of non-affordable luxury clothing fashion is De Grisogono.
- affordable luxury business: the products quality is satisfactory and they are still stylish, but their price is at the level which makes them available for a wider group of customers. The products brand gives the customers the social status and makes them feel belonging to a certain social group (Kapferer and Bastien 2009, 2012). Example of affordable clothing luxury fashion is Missoni and Sector.

Regarding luxury business, an important issue constitutes analysis of the innovation strategy, whose aim is to ensure the survival of the company in a long-term period (Giacosa et al. 2014; Re 2013).

According to a widespread literature, products innovation has been distinguished from processes innovation (O'Donnell et al. 2002; Sood and Tellis 2009; Utterback and Abernathy 1975). Such a division provides relevant theoretical and practical implications for the innovation strategy (Adner and Levinthal 2001; Bonannoa and Haworthb 1998; Drejer 2004). The division mentioned above is not always clear-cut (Giacosa 2015); however, it is observed that the combination of both of them results in improvement of the company's performance (Roberts 1988).

In the context of products innovation in the luxury business, it relates to development of new products or to improvement of the relevant features of already existing items. In this way consumer can enjoy an increased utility of the product

(Giacosa 2011; Re et al. 2014; Sabisch 1991), keeping its loyalty to a certain brand (De Chernatony 2001; Okonkwo 2007; Ross and Harradine 2011; Vaid 2003). A process innovation instead relates to development of new operating methods, improvement of existing ones or to use the production factor in a different way increasing efficiency of the production in terms of costs, quality and service (Giacosa 2011).

Present publication is therefore aimed at filling the gap in the literature on the innovative luxury business models and to verify if development and adaptation of innovative solutions can help the company to achieve a competitive advantage.

3 Methodology

The main objective of this research is to verify how the business model could increase the competitiveness of a luxury company. In our research, we referred to Giacosa (2016) which analysed how the investments strategy implemented by a luxury family firm could represent a mean of sustainable development in terms of economic, social and cultural aspects. Indeed, thanks to its history and reputation, the family firm could increase the integrity of its business, generating a positive impact on its turnover, on the employment and, in general, on the society.

This research is placed in a wider context and it is not linked only to family business, but it refers to the luxury business (in which both family and non family businesses operate). It was sought to do a parallelism as a family business that operates in luxury sector often takes advantage of familiness factor (Arregle et al. 2007; Frank et al. 2010; Irava and Moores 2010; Minichilli et al. 2010; Sirmon and Hitt 2003) in order to increase its long-term competitiveness. In this way a luxury company could exploit innovation factor as the element of evaluating its business model. The methodology of present research has been conducted in the following phases:

- the first phase: a literature review on the luxury business, with the purpose to analyse different categories of luxury products and make a complete contextual framework. It allows to qualify the type of offer and consequently the typologies of businesses. In addition, due to the purpose of the research, we focused on the innovation strategy in the luxury business, as it permits to understand the role of investments strategy in the development of the competitive advantage;
- the second phase: the treatise was conducted using qualitative approach. Conducting current analysis was possible thanks to the observation of different globalized and innovative luxury medium-sized companies. In particular, we identified a series of characteristics of luxury innovative companies to ensure a long-term competitive business model.

Despite its criticism, the qualitative model has been applied as it enables in-depth understanding of the luxury companies' business model, and avoids a descriptive approach. What is important is that the effectiveness of the qualitative methods can be adopted in the luxury context (George and Bennett 2005; McKeown 2004; Yin 2009).

We used the Schematic view (Lau and Woodman 1995) for identifying specific business model's features, as it permits to identify a set of factors that are relevant in facilitating and supporting the decision-making process carried out by the company's management and/or ownership.

4 The Business Model's Features

Companies operating in the luxury business have to adopt an effective business model to increase their competitive advantage, especially due to a series of paradigms such as the consumption crisis, the luxury democratization and the market globalization. To reach this purpose, the company has to adopt an effective and efficient innovation strategy which is influenced by several variables.

Using the Schematic view, we referred to our framework (Giacosa 2015), which introduced a set of factors that are relevant in facilitating and supporting the decision-making process carried out by the company's management taking into account the investments strategy. In particular, the main critical factors thanks to which the business model proposed should be articulated are the following:

- Innovativeness;
- Company assets;
- Long-term vision in the investments attitude.

(A) **Innovativeness**

- With reference to innovativeness, it should consider productive, organizational and commercial aspects presented below (Giacosa 2012):
- strict quality control can lead to higher quality of the products (O'Donnell et al. 2002) and reduction of production costs. It permits to reduce the manufacturing faults and improve characteristic of the products. As an example Missoni company can be presented, which thanks to adopting innovative processes is able to increase the products quality and their attractiveness, maintaining in the same time its core business;
- attracting attention of new and old customers requires consideration of the features such as shops location, their type and outlook, as well as development of new marketing channels. It is essential to introduce some new instruments aimed at catching the consumers attention and ensuring competitive advantage. In this context, multi-sensorial tools (Kapferer 2002; Ward 1997) and new digital and innovative instruments (Donaldson 2011; Geerts and Veg-Sala 2011; Kim and Ko 2012; Okonkwo 2009) can be used. They contribute to the products development, encouraging the customers interest and refreshing the company's brand;

– improvement of the organizational structure of innovation strategy requires an intervention of human resources and any other company's functional area in order to suit to the needs of innovation strategy.

Lack of the features described above is connected with a risk of limited vision of innovation. As a result, innovation strategy can be considered as limited investment that do not involve the company as integrity (Cantino 2007).

Innovativeness can be divided in two groups (Giacosa 2015): incremental or radical innovation. The choice depends on the company's aversion to risk. Nevertheless, it is important to impose the innovativeness only at innovative costs to reach economic benefits and keep the customers loyalty. When introducing of innovativeness does not bring commercial results, future survival of the company becomes the most important issue which is possible to achieve by combination of creative activity and commercial venture (Giacosa 2012). It can result in a negative performance of innovative products. On the other hand, an excessive innovation can cause troubles, as well. If the products created by the company constitute a copy of the items available on the market, the customers can evaluate it in a negative way and lost an interest in that product. As a result, it can lead to deterioration of both, the company's turnover (in a short-term) and commercial image of the company (in a long-term). It is therefore important to be aware of the risks, because ineffective innovation and lack of innovativeness can have a negative impact on the competitiveness of the company and business sustainability.

The innovativeness factor is described below (Fig. 1).

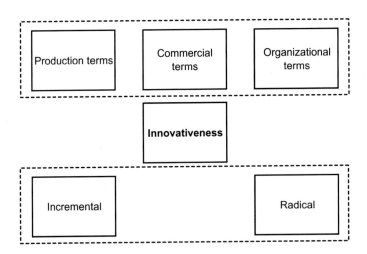

Fig. 1 The innovativeness factor. *Source* personal elaboration

(B) The Company's Assets

Innovation strategy should take into account the company's assets (Bordieu 1977; Davis 1983) which reflects the company's heritage accumulated over generations (in terms of reputation, trust and values). It is very essential as it ensures an authenticity of the company's brand in a long-term vision (Eckrich and Loughead 1996; Lambrech 2005; Perricone et al. 2011). For instance, for Brunello Cucinelli the company's assets value refers to the philosophy of the company across the generations.

The issue of corporate assets values becames so relevant as the company's heritage has a positive impact on the prestige of the brand (Attanzio 2011; Ferda 2010), meaning that the quality should be coherent with the brand value, as the quality factor has a significant influence on the customer's assessment.

It is important to note that the company should pay attention to details connected with production and management of relationships with customers. They should be coherent with the company's philosophy and values as it allows to maintain the customers loyalty (Bresciani et al. 2013).

Another factor that impacts the patrimony of the company is price policy, which depends on the availability of the luxurious goods. It is observed that in case of non-affordable luxury products the price remains at the same high level regardless even the sales seasons (Everyday High Pricing policy) It is believed that reduction of prices could have negative influence on the brand's perception and customer could lost its interest in such a product. On the other hand, companies selling affordable luxury products use High/Low Pricing policy—they keep high prices during the season and take into account reduction of prices when sales season starts. In this case it is relevant to increase the sells threshold in order to improve the business performance.

In addition, it is essential that the innovation strategy is coherent with company's identity and its heritage (Kapferer 2008). Otherwise, the prestige which the company has established over the years can suffer a deterioration and the business can become non-sustainable.

Regarding specifically non-affordable luxury, an important factor that attract the customers interest is a rarity of a product. In this context, deterioration of the corporate assets value could affect in negative way the rarity of the products and consequently, reduce the feeling of luxury and prestige. For example, the Kelly brand is characterised by limited availability or long waiting periods what makes it even more exclusive. Therefore, the corporate assets value should be in line with the exclusivity of the brand and the perception of the customer's social status.

It turned out that the company should always consider needs of all customers (Moody et al. 2010). In this way the innovation strategy permits to satisfy the customer's needs as well as his perception of belonging to a certain social class. In this context, the business becomes sustainable also regarding the satisfaction of the consumers.

The company's assets factor is described below (Fig. 2).

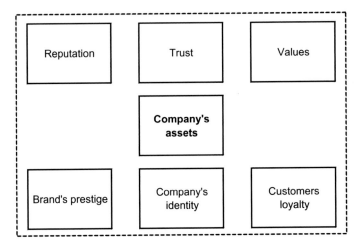

Fig. 2 The company's assets factor. *Source* personal elaboration

(C) The Long-term Vision in the Investments Attitude

Consideration of the long-term vision is necessary due to the transfer of the company to the future generations (Neumann 1997; Ward 1997; Zellweger 2007).

It defines a choice of investments and development of the most appropriate conditions to generate the value over time (Davis 1983; Habbershon and Pistrui 2002; Naldi et al. 2007; Zahra 2005; Zahra et al. 2004).

When there is no perspective of continuation the company's activity in the future, the attitude of the entrepreneur or management to a company is less strong (James 1999). In this context, a survival of the company and maximizing the return on the company's social and symbolic assets becomes a priority for the company's shareholders'. In this situation (lack of the company's long-term vision) the competitive in luxury sector is hard, as creation of the brand and extending of a marketability constitute a long-lasting processes.

In a short-term strategy, the creation and valorization of a brand is impossible, what results in deterioration of the company's prestige and luxurious status in a long-term period.

In summary, an incorrect innovation strategy in long-term can make the business non-innovative and as a result, can destroy the company's image and harm the heritage which a luxury brand should conserve (Fig. 3).

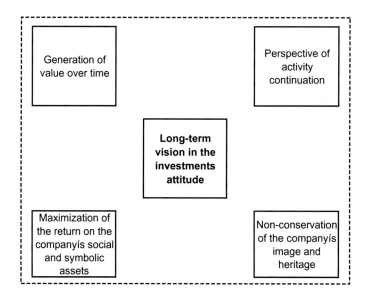

Fig. 3 The long-term vision in the investments attitude factor. *Source* personal elaboration

5 Conclusions, Implications and Limitations

In many countries, such as Italy, the luxury business is one of the most important industries which is characterized by significant growth, development and economic sustainability. In the recent years, it is affected by some paradigms (such as consumption crisis, democratization process, market globalization) that should be considered while defining the company business model.

In particular, in case of non-affordable luxury products it is essential to increase the intrinsic value of the products, making them available just for a certain group of customers (Kapferer and Bastien 2009; Giacosa 2012). With reference to affordable luxury products the objective is to increase the brands perception and strengthening the customers loyalty what is possible by introducing creative and innovation activities.

To overcome the above paradigms, the business model has to be characterized by a set of factors which could increase the competitiveness of a luxury company. In particular, the following key drivers have been distinguished:

- Innovativeness;
- Company assets;
- Long-term vision in the investment attitude.

Regarding innovativeness it should consider productive, organizational and commercial aspects such as quality control, attracting attention of new and old customers and improvement of organizational structure of the innovation strategy. Lack of these features is connected with a risk of limited vision of innovation.

Moreover, a company has to be aware of the fact that ineffective innovation and lack of innovativeness can have a negative impact on the competitiveness of the company and business sustainability.

With reference to company assets, the innovation strategy should take them into account seeing that they reflect the company's heritage accumulated over generations (especially in terms of reputation, trust and values). It is very important because it ensures an authenticity of the company's brand in a long-term period.

Finally, consideration of the long-term vision is necessary due to the transfer of the company to the future generations. The incorrect innovation strategy in long-term vision can make the business non-innovative and, as a result, can destroy the company's image and harm the heritage that a company has established over the years and which a luxury brand should conserve.

It is important to highlight that the key factors described above should be applied together to understand better the phenomenon of innovative luxury business. Consideration of all of them ensures the improvement of the company's business model and making the model more competitive in the international context (Fig. 4). Moreover, to ensure the functionality of the system, the relation between the components of the company's system should be maintained.

Current research contributes with the following theoretical an practical implications:

- regarding theoretical implications—they relate to the potential of some business model's features for a luxury sector. One of its advantages is more efficient management of the investment strategies. Present research analysed the relation and impact of the key factors on the business model developed in the context

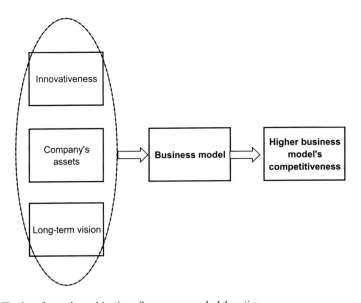

Fig. 4 The key factors' combination. *Source* personal elaboration

analysed. Only combination of the factors introduced permits to understand the influence of investment strategy on the promotion of innovative business;
– regarding practical implications—they relate to the luxury context which is interesting field of opportunities for companies in many typologies of sub-context. Moreover, some characteristics of the luxury business reflect competitive advantage as they define better the management of the investment strategy.

Current research has also recorded some limitations. Its extension about comparison of potential dynamics of various luxury sectors would constitue an interesting contribution. Moreover, to simulate better the impact of key factors on the benefits of investment strategy, the suitable econometric models should be used.

The future researches on this issue should consider also this aspect: nowadays, when the economic crisis caused the decrease in consumption, the luxury sector has recorded opposite tendency. Future researches should therefore focus on the ability of gaining financial sources from abroad to invest in innovation strategy. It would enable the companies to face and overcome the difficulties.

References

Adner R, Levinthal D (2001) Technology evolution and demand heterogeneity: implications for product and process innovation. Manage Sci 47(5):611–628

Arregle JL, Hitt MA, Sirmon DG, Very P (2007) The development of organizational social capital: attributes of family firms. J Manage Stud 44(1):72–95

Attanzio G (2011) Le imprese familiari resistono meglio alla crisi economica. http://www. impresanews.it (home page). Accessed 4 Aug 2014

Bonannoa G, Haworthb B (1998) Intensity of competition and the choice between product and process innovation. Int J Ind Organ 16(4):495–510

Bordieu P (1977) Outline of a theory of practice. University Press, Cambridge

Bresciani S, Bertoldi B, Giachino C, Ferraris A (2013) The approach of family businesses in the luxury industry. In: Conference readings book proceedings, in 6th euromed conference of the euromed academy of business. Confronting contemporary business challenges through management innovation, Estoril, Portugal, 3–4 Oct, pp 423–433

Cantino V (2007) Corporate governance, misurazione della performance e compliance del sistema di controllo interno. Giuffrè, Milano

Chevalier M, Mazzalovo G (2008) Luxury brand management. Franco Angeli, Milano

Corbellini E, Saviolo S (2009) Managing fashion and luxury companies. Etas, Milano

Davis JH (1983) Realizing the potential of the family business. Org Dyn 12(1):47–56

De Chernatony L (2001) From brand vision to brand evaluation strategically building and sustaining brands. Butterworth Heinemann, Oxford

Donaldson MS (2011) Promoting luxury goods in china through social media. MultiLingual (Oct/Nov), pp 29–31

Drejer I (2004) Identifying innovation in surveys of services: a Schumpeterian perspective. Res Policy 33(3):551–562

Eckrich CJ, Loughead TA (1996) Effects of family business membership and psychological separation on the career development of late adolescents. Family Bus Rev 9(4):369–386

Ferda E (2010) Family business reputation: a literature review and some research questions electronic. J Family Bus Stud 4(2):133–146

Frank H, Lueger M, Nosé L, Suchy D (2010) The concept of 'Familiness'. Literature review and systems theory-based reflections. J Family Bus Strategy 1(3):119–130

Gardetti MA (2011) Sustainable luxury in Latin America. In: Conference dictated at the seminar sustainable luxury & design within the framework of the MBA of IE Business School, Madrid, Spain

Geerts A, Veg-Sala N (2011) Evidence on internet communication—management strategies for luxury brands. Global J Bus Res 5(5):81–94

George A, Bennett A (2005) Case studies and theory development in the social sciences. MIT Press, Cambridge

Giacosa E (2011) L'economia della aziende di abbigliamento. Giappichelli, Torino

Giacosa E (2012) Mergers and acquisitions (M&As) in the luxury business. McGraw-Hill, Milano

Giacosa E (2015) Innovation in luxury fashion businesses as a means for the regional development. In: Cagica Carvalho Luísa (ed) Handbook of research on entrepreneurial success and its impact on regional development, pp 206–222, IGI Global Publisher, Hershey

Giacosa (2016) Family business phenomenon luxury business clothing luxury business democratization sustainable luxury business. In: Gardetti MA (eds) Sustainable management of luxury, pp 361–385, Springer

Giacosa E, Giachino C, Bertoldi B, Stupino M (2014) Innovativeness of ceretto aziende vitivinicole: a first investigation into a wine company. Int Food Agribusiness Manage Rev 17 (4):223–236

Habbershon TG, Pistrui J (2002) Enterprising families domain: family influenced ownership groups in pursuit of trans generational wealth. Family Bus Rev 15(3):223–238

Irava WJ, Moores K (2010) Clarifying the strategic advantage of familiness: unbundling its dimensions and highlighting its paradoxes. J Family Bus Strategy 1(3):131–144

Jackson TB (2004) International retail marketing. Elsevier Butterworth-Heinemann, Oxford

James H (1999) Owner as manager, extended horizons and the family firm. Int J Econ Bus 6 (1):41–55

Kapferer JN (2002) Ce qui va changer les marques. Editions d'Organisation, Paris

Kapferer JN (2008) The new strategic brand management. Kogan Page, London and Philadelphia

Kapferer JN, Bastien V (2009) Luxury strategy. FrancoAngeli, Milano

Kapferer JN, Bastien V (2012) The luxury strategy: break the rules of marketing to build luxury brands. Kogan Page Limited, London

Kim K, Ko E (2012) Do social media marketing activities enhance customer equity? An empirical study of luxury fashion brand. J Bus Res 65(10):1480–1486

Kleanthous A (2011) Simple the best is no longer simple raconteur on sustainable luxury. http://www.theraconteurcouk/category/sustainability/sustainable-luxury.html. Accessed Dec 2012

Lambrech J (2005) Multigenerational transition in family businesses: a new explanatory model. Family Bus Rev 18(4):267–365

Lau C, Woodman RC (1995) Understanding organizational change: a schematic perspective. Acad Manag J 38(2):537–554

McKeown T (2004) Case studies and the limits of the quantitative worldview. In: Brady H, Collier D (eds) Rethinking social inquiry. Lanham MD, Rowman and Littlefield

Miller KW, Mills MK (2012) Contributing clarity by examining brand luxury in the fashion market. J Bus Res 65(10):1471–1479

Minichilli A, Corbetta G, MacMillan IC (2010) Top management teams in family controlled companies: familiness, faultlines, and their impact on financial performance. J Manage Stud 47 (2):205–222

Moody W, Kinderman P, Sinha P (2010) An exploratory study: relationships between trying on clothing, mood, emotion, personality and clothing preference. J Fashion Mark Manage 14 (1):161–179

Naldi L, Nordqvist M, Sjoberg K, Wiklund J (2007) Entrepreneurial orientation, risk taking, and performance in family firms. Family Bus Rev 20(1):33–47

Neumann M (1997) The rise and fall of the wealth of nations: long waves in economics and international politics. Edward Elgar, Cheltenham and Lyme

O'Donnell A, Gilmore A, Carson D, Cummins D (2002) Competitive advantage in small to medium sized enterprises. J Strateg Mark 10(3):205–223

Okonkwo U (2007) Luxury fashion branding. Palgrave Macmillan, New York

Okonkwo U (2009) The luxury brand strategy challenge. J Brand Manage 16(5/6):287–289

Perricone PJ, Earle JR, Taplin IM (2011) Patterns of succession and continuity in family-owned businesses: study of an ethnic community. Family Bus Rev 14(2):105–121

Re P (2013) La gestione dell'innovazione nelle aziende familiari. Giappichelli, Torino

Re P, Giacosa E, Giachino C, Stupino M (2014) The management of innovation in the wine business. 2nd International symposium systems thinking for a sustainable economy advancements in economic and managerial theory and practice conference proceedings, January 23–24. Roma, Italy, pp 1–17

Roberts EB (1988) Managing invention and innovation. Res Tech Manage 31(1):11–29

Ross J, Harradine R (2011) Fashion value brands: the relationship between identity and image. J Fashion Mark Manage 15(3):306–325

Sabisch H (1991) Product innovation. C.E. Poeschel, Stuttgart

Shipman A (2004) Lauding the leisure class: symbolic content and conspicuous consumption. Rev Social Econ 62(3):277–289

Silverstein MJ, Fiske N (2003) Luxury for the masses. Harvard Bus Rev 81(4):48–57

Sirmon DG, Hitt MA (2003) Managing resources: linking unique resources, management, and wealth creation in family firms. Entrepreneurship: Theory Pract 27(4):339–358

Sood A, Tellis GJ (2009) Do innovations really payoff? Total stock market returns to innovation. Mark Sci 28(3):442–456

Thomas D (2007) Deluxe—how luxury lost its luster. Penguin Books, New York

Utterback JM, Abernathy WJ (1975) A dynamic model of process and product innovation. Omega 3(6):639–656

Vaid H (2003) Branding. Cassel Illustrated, London

Ward JL (1997) Growing the family business: special challenges and best practices. Family Bus Rev 10(4):323–337

Yin RK (2009) Study research: design and methods, 4th edn. Sage Publications, Thousand Oaks

Zahra SA (2005) Entrepreneurial risk taking in family firms. Family Bus Rev 18(1):23–40

Zahra SA, Hayton JC, Salvato C (2004) Entrepreneurship in family vs non-family firms: a resource based analysis of the effect of organizational culture. Entrepreneurship Theory Pract 28 (4):363–381

Zellweger T (2007) Time horizon, costs of equity capital, and generic investment strategies of firms. Family Bus Rev 20(1):1–15

Appreciative Mentoring as an Innovative Micro-Practice of Employee Engagement for Sustainability: A Luxury Hotel's Entrepreneurial Journey

Gulen Hashmi

Abstract A significant dilemma for any business in today's sustainability age is the ability to acknowledge society's challenges and create innovations to address those challenges. This necessitates a change of perspective in the approach to organizational transformation strategies, coupled with the need to establish a strong sustainability culture where superior levels of organizational engagement prevail. The focus of this article is to showcase how, within a strong organizational culture of sustainability, appreciative mentoring can be used as an innovative entrepreneurial micro-practice to facilitate positive conversations and engagement in the organizational transformation process of enhancing employee engagement. The research in this article was carried out in two phases as a sequential transformative mixed methods design in an action research (AR) study, to support the use of a transformative perspective in research and ultimately help organizations to address employee engagement with a participative and democratic orientation to knowledge creation. Our research consists of a sustainability culture and leadership assessment survey, and subsequent appreciative mentoring conversations co-facilitated with a luxury London hotel leading in sustainability. The research sets the stage for further empirical research to determine the impact of appreciative mentoring on the mentor-mentee relationship and the organization's engagement level over time.

Keywords Appreciative inquiry (AI) · Appreciative mentoring
Mentor-mentee relationship · Organizational sustainability · Innovative
micro-practice · Entrepreneurial capability · Organizational culture of sustainability
Organizational transformation · Sequential transformative mixed methods design
for action research

G. Hashmi (✉)
Business School Lausanne, Lausanne, Switzerland
e-mail: gulen.hashmi@bsl-lausanne.ch

© Springer Nature Singapore Pte Ltd. 2018   31
M. A. Gardetti and S. S. Muthu (eds.), *Sustainable Luxury, Entrepreneurship,
and Innovation*, Environmental Footprints and Eco-design of Products
and Processes, https://doi.org/10.1007/978-981-10-6716-7_3

1   Introduction

In the face of troubling scientific facts and figures about the unsustainable state of our economy and environment (Hashmi and Muff 2015), a major issue for many of today's organizations is the dependency of their sustainability on changes such as increased innovation or the fight to secure increasingly rare and demanding employees (Cooperrider and Fry 2013). Advanced technologies, which are the cornerstones of business development in today's world, are dependent on skilled human resources. As the futurist Naisbitt has said, it is not advanced technology but high quality people (hi-touch) that best promotes development (Naisbitt 2001). Yet, qualified individuals with a committed and engaged outlook on work can only thrive in a positive and constructive organizational culture (Baltas 2013). As such, we could easily make the analogy that sustainability performance is very much like the relationship between seed and soil where employees are the seeds and the organizational culture is the soil wherein the seed can bear fruit.

In the context of organizational transformation for sustainability, organizational members—both employees and line managers—play an important role in determining the organizational change process, and whether an organizational intervention is successful in improving employee engagement and well-being (Nielsen and Randall 2012). Entrepreneurial capabilities, which are defined as the ability to perceive, choose, shape and synchronize internal and external conditions for the enterprises' exploration and exploitation (Zahra 2011), determine employees' engagement to a change process or initiative, which consequently leads to the success or failure of corporate change strategies (Sekerka et al. 2009). In addition, employee engagement leads to innovative behavior where employee goes beyond individual roles to collaborate with colleagues, make suggestions to improve the business environment, and work to better his or her company's standing in the external environment (Sundaray 2011). The Gallup Organization, potentially the most widely recognized name associated with employee engagement, indicates that there is a direct cause and effect relationship between employee engagement and organizational innovation (Gallup 2015).

Studies have found that when an employee's values fit the organization's values, the employee will stay longer and be more engaged and productive (Kristof-Brown et al. 2005). Thus, helping the employees to create and develop a positive and constructive organizational culture aligned with their values would serve the organization's long-term sustainability goals as well as contribute positively to employee engagement strategies. Indeed, companies need to first develop a strong culture of sustainability when moving towards profound change related to business sustainability (Crane 1995). In this paper, we consider a strong culture of sustainability to be one in which organizational members hold shared assumptions, values and beliefs about the importance of balancing economic efficiency, social equity and environmental accountability (NBS 2012). To illuminate how the employee engagement of a luxury London hotel leading in sustainability might be transformed and enhanced, we focused on the following two primary questions, which were explored in subsequent phases of our research, respectively:

1. What are the differences in employees' perceptions of an organizational culture of sustainability across the ranks of the hotel, which justify the need for an action research intervention on improving employee engagement at the hotel?
2. How can this luxury hotel move to a stronger culture of sustainability and enhance employee engagement by leveraging the potential of appreciative mentoring conversations?

In the first phase, we focused on understanding and describing the status of certain aspects of the hotel's existing organizational culture of sustainability, specifically with regard to differences in perceptions of various staff members from different departments of the hotel. Using a Sustainability Culture and Leadership Assessment (SCALA)[1] survey, we studied the culture of sustainability of the hotel with regard to a broad range of organizational aspects, such as organizational leadership, organizational systems, organizational climate, change readiness including entrepreneurial capabilities, internal stakeholders and external stakeholders. These organizational aspects helped us better assess the potential impact of an AR intervention on employee engagement that we planned for the second phase of our research. The AR intervention consisted of five pilot appreciative mentoring conversations, the focus of which were to create and share knowledge on how the hotel could enhance employee engagement for sustainability in a participatory, empowering and entrepreneurial way through action and reflection.

The AR intervention was perfectly aligned with our transformative theoretical perspective—the Appreciative Inquiry (AI) framework and the ideology of empowerment. AI was developed as an affirmative form of knowledge creation that would serve as a pathway to social innovation (Cooperrider and Srivastva 1987, p. 159). Conceptualized as a second-generation form of AR (Cooperrider and Godwin 2012), AI is meant to be an inquiring approach to organizational development and change. It is *not* about implementing a change to get somewhere; it *is* about changing, convening, conversing and relating with each another in order to tap into the natural capacity for collaboration (Bushe and Coetzer 1995) and entrepreneurial change that is in every system. AI is a reconfiguration of AR that is a collaborative democratic partnership among the members of a system (Coghlan and Brannick 2010). In AR, knowledge is generated both with various stakeholders and within each participant as a co-researcher, and each participant is engaged to understand a concept or practice more deeply, through a systematic empirical inquiry.

In this article, we therefore seek to take the call for systematic empirical inquiry to a new level by collectively exploring the appreciative interview process within

---

[1]The Sustainability Culture and Leadership Assessment (SCALA) survey is an assessment instrument composed of items pertaining to culture and leadership. The assessment contains both sustainability-specific content as well as more general organizational climate content that has been demonstrated or asserted in other research to impact the execution of sustainability strategy.

the micro-practice of mentoring conversations—as small group, face-to-face entrepreneurial interactions—rather than within macro, large-scale social processes at the inter-organizational, the state and national levels. We do this by emphasizing the agency of various operational staff and mentors through a focus on studying subjective interpretations and perspectives, rather than seeking to pose structural explanations for large-scale patterns and trends at the macro level. Indeed, research suggest that as much or more change emerges from daily interactions at work—as people discuss the inquiry, exchange stories, and are impacted by new conversations—as it does from new ideas or plans (Kessler 2013). Similarly, Whitney and Trosten-Bloom (2003, p. 161) have found that the one-to-one, face-to-face appreciative interview seems to stimulate the most energy and enthusiasm among cynical and inspired, young and old, professional and frontline alike.

As systemic action is the primary leverage point for successful change (Chouinard et al. 2011), mentoring one-on-one dialogues can serve as an ideal platform where generative AI conversations can take place as systemic actions in organizations. Mentoring is a learning relationship that has widely been accepted (Hansman 2002) as a positive method to promote organizational learning and entrepreneurial initiatives, help new employees integrate to the workplace culture, enhance leadership skills essential to leading sustainability initiatives, provide developmental and psychological support and enhance engagement for sustainability (Higgins and Kram 2001). The micro dynamics of the mentor-mentee relationship are sensitive to the larger organizations in which they reside; therefore, they are "influenced by the macro dynamics of intergroup power relationships in organizations…resulting in subtle or dramatic shifts in power relations among groups in organizations" (Ragins 1997).

In our study, we therefore, sought to bring the notion of "non-hierarchical, democratic relationships" at play (Darwin 2000) through appreciative mentoring conversations that were initiated to run as a mentor-driven, entrepreneurial activity, with the aim of enhancing employee engagement of a luxury hotel leading in sustainability.

2   Literature Review

As part of the preparation for the research presented in this paper, we first carried out a literature review on organizational culture as it relates to sustainability implementation. Having only taken account of empirically tested conclusions in the literature, we found that organizational culture can be a useful tool for organizational change programs related to embedding sustainability as the values and ideological underpinnings of a company's culture affect how sustainability is implemented (Cameron and Quinn 2006; Jarnagin and Slocum 2007). Culture is often described as "the way we do things around here" (NBS 2012). Organizational culture includes norms and values regarding appropriate and desirable interactions with others both outside and within the walls of the organization (Miller Perkins

2011). If a more workable definition is required, we may describe organizational culture as what employees do when they are not assigned a task and are not being monitored. Collins (2001) posits that, although their numbers are scarce, companies that achieve success in their sustainability strategic integration are not averse to change and have enduring values.

Most sustainability-related surveys conducted over the past few years have acknowledged that culture plays an important role in the implementation of sustainability strategies (Miller-Perkins 2011). Glisson (2007) has defined a positive organizational culture as one that possesses positive staff morale, reduced staff turnover, greater sustainability of existing and future sustainability initiatives, and improved service quality. Though there are many features of a positive organizational culture, the following are noted as the most salient ones by Armstrong (2010): employee empowerment and effective communication, open and honest communication, long-term business excellence, individual responsibility, adoption of new ideas, and flexible and rapid responsiveness. Woodman et al. (1993) rightly argued that organizational innovation is dependent on the creativity of the organizational culture, which in turn is dependent on individual creativity. Innovations reflect the creative efforts of employees who build and promote an innovative culture.

Harter et al. (2002), in a meta-analysis they conducted, found that leadership style and organizational climate have significant effects on employee engagement and strategy execution. Employee engagement feeds on trust and innovation. According to Unsworth (2003), innovation is a process of engaging in behaviors designed to generate and implement new ideas, processes, products and services. However, behavioral thinkers define innovation as a mindset, which is influenced by beliefs, values and behavior. Implementation mechanisms such as enterprise-wide management systems or performance evaluation and compensation processes—that require employee engagement—ensure that change happens as innovations diffuse throughout organizations. In turn, a culture supportive of sustainability will increase the effectiveness of leadership commitment, external engagement, employee engagement and mechanisms for implementation (Eccles et al. 2012).

Employee engagement is widely accepted by practitioners and academics as it has a significant and positive impact on both the individual and organization (Schaufeli and Bakker 2010). Engagement experts state that engagement is key to innovation and competitiveness (Gichohi 2014). Yet, there is no general consensus on the conceptualization of employee engagement. The majority of academics define engagement as a psychological state whereby engagement is seen, as Schaufeli and Bakker (2010) define it, as the antithesis of burnout, characterized by vigor, dedication and absorption (fully concentrated in one's work). As for the business side, the majority of the management consultancies and HR professionals define employee engagement in terms of organizational commitment (a desire to stay with the organization in the future) and employees' willingness to 'go the extra mile', which includes discretionary effort in effective corporate functioning (CIPD 2011). For the AR research presented in this paper, however, the following Lewis et al. (2011) definition of employee engagement is preferred as it encompasses the range of definitions across both academic research and practice:

> Being focused in what you do (thinking), feeling good about yourself in your role and the organization (feeling), and acting in a way that demonstrates commitment and alignment to the organizational values and objectives (acting).

Bushe and Kassam (2005, p. 165) found that transformational organizational change was associated with an improvisational, rather than a planned approach to change that is dictated by organizational rules and guidelines. In this work, improvisation is basically considered as "the deliberate and substantive fusion of the design and execution of a novel production" following the definition given by Miner et al. (2001: 314). Storytelling, by nature, possesses such an improvisational approach to change since it "expresses how and why life changes," which lowers emotional resistance to change by getting to the "heart" (Mckee 2003). In this regard, AI is a practical AR methodology that emphasizes the conversational and sense-making life of the organization as sources of entrepreneurial capability and change. AI contributes to development of entrepreneurial capabilities through exploration—recognition, discovery and dream of the positive core—as well as exploitation—developing opportunities through an ongoing understanding of discovery and learning.

Since the early 1980s, AI has been increasingly used by thousands of people and hundreds of organizations in every sector of society to promote transformative change (Cooperrider et al. 2005; Thatchenkery and Metzker 2006). Barrett and Fry (2005), for instance, describe AI as a process of building cooperative capacity for excellence and innovation. The power of AI is the way in which organizational members become engaged and inspired by focusing on their own positive experiences through the use of AI approaches (Bushe 2001). This can be accomplished in a day or require multiple-day summits or retreats. The AI process can be as formal as a yearlong, whole system process, and one that involves a broad range of internal and external stakeholders in the change process. Such processes can take the form of large AI summits that are single events or series of events that bring people together; or it may be as informal as a conversation between a manager and a subordinate (Cooperrider et al. 2008, p. 101).

AI is a highly psychological approach in that it focuses on the social-political aspects of organizational life using the power of positive emotions. In empirical research, positive emotional states have been observed as associated with more socially oriented behavior, greater curiosity and entrepreneurial capabilities, and greater willingness to accommodate ambiguity and uncertainty inherent in change (Fredrickson and Branigan 2005). The studies show that AI can lead to meaningful change in organizations, and that its strength-based inquiries can influence employees' readiness for change and overcome negative reactions to upcoming change (Sekera et al. 2006; Sekerka et al. 2009). As such, the AI process is built upon an initial inquiry that asks questions about values, successes and strengths, which is in itself transformational, based on the premise that "organizations move toward what they study" (Cooperrider et al. 2003, p. 29).

In Appreciative Inquiry, this initial inquiry, or *Discovery*, is part of a what is termed a 4D Cycle—the most often used depiction of the AI process offered by Cooperrider and Srivastva (1987)—that leads groups/individuals through a step wise process to later *Dream* about what could be, *Design* a future, and take action to change their *Destiny*. The "4D's" of the AI process represent different, intentional sets of activities and conversations, all linked to an affirmative inquiry topic for *Discovery*. The list of affirmative/positive topics for discovery is endless: integrity, high quality, empowerment, innovation, team spirit, customer responsiveness, and so on. However, the linearity of the AI process should not be mistaken for a "forced march" agenda that one must follow, as after *Discovery*, AI processes can take varied paths. Below is a depiction of the steps of the AI process, which involves the co-operative search for the best in people, their organizations, and the world around them (Fig. 1):

The *Discovery* phase is a time for diverse members of the organization or community to share positive experiences. Moreover, participants discuss in depth the organization's positive core (Cooperrider et al. 2008). Facilitation of this phase often includes appreciative one-on-one interviews with staff by consultants or the staff themselves. Appreciative interviews offer every participant a chance to discover their own thinking in the relative safety of a one-on-one dialogue, and quickly generate a deep sense of interconnectedness among pairs (Ludema and Fry 2008). According to Cooperrider and Whitney (1999, p. 11), *"At AI's heart is the appreciative interview. The uniqueness and power of an AI interview stems from its fundamentally affirmative focus"*. During the appreciative interview, staff members are asked to share positive and meaningful experiences from working in their organizations. Interviews vary in length (from 20 to 90 min), depending on the circumstances of the intervention. Highlights from the interviews are then shared in larger groups in the form of feedback sessions.

The *Dream* phase allows participants to envision their organization with a foundation built on the exceptional and positive experiences discussed in the previous phase. In this phase, participant stories help co-construct a vision of the organization's positive impact. The *Dream* phase is generally conducted in small groups working to envision their organization's potential.

Fig. 1 Appreciative Inquiry 4D model

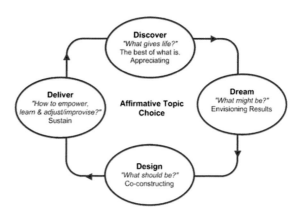

The *Design* phase articulates the systems and structures necessary to support positive experiences and co-created dreams. Participants have the opportunity to co-create design statements that highlight the positive qualities and realities discussed in the previous two phases (Cooperrider et al. 2003, p. 24). In this phase, basic project plans begin to form and participants feel empowered to take action.

Finally, the *Destiny* phase involves language that relates to practical implementation of those systems. Participants create a list of inspired action-oriented tasks that reflect the work of previous phases (Cooperrider and Whitney 2005). Selection of possible action tasks and the formation of task forces that discuss and establish principles for future work is a key activity in this phase. The *Destiny* phase of AI is an ongoing understanding of the *Discovery*, *Dream* and *Design* phases. It is an iterative cycle of inquiry with discovering and learning continuously occurring. Through such inquiry participants construct new perceptions of reality, based on the assumption that inquiry is already an intervention in itself (Barrett and Fry 2005, p. 46).

The theory of a socially constructed reality or social constructionism, is key to AI because it shows that language is not the sole means for exchange of meaning, but that language itself plays an active role in the creation of meaning (Busche 2005, p. 123). Social constructionism is a philosophy of science that believes individuals have a considerable influence over the reality that they perceive and experience; moreover, individuals generally create their own reality through collective symbolic and mental processes (Cooperrider et al. 2008). This implies that language that shares and discovers new ideas or visions can also be a powerful tool for change.

Interestingly, mentoring relationships are socially constructed power relationships, and the power that mentors have and exercise within mentoring relationships can be helpful and engaging, or hurtful and disengaging to mentees. Johnson-Bailey and Cervero (2001, 2002) discuss mentoring as occurring on two dimensions: the internal dimension, which is the relationship between the mentor and mentee, and the external context that encompasses the mentoring pair and the organization. In spite of good organizational intentions, however, many mentoring programs planned by organizations are unsuccessful and fail to remove barriers to advancement for mentees (Thomas 2001), reflecting the power and interests prevalent in many organizations. Thus, organizational interests may be served at the cost of employee interests (Hansman 2000; Bierema 2000). As traditional mentoring relationships are hierarchical, composed of one experienced person who advises a less experienced person (Brinson and Kottler 1993), the mentor's role to "fill up the mentee with knowledge" denies the validity of the ontological and epistemological stances of the learner.

Thus, in our research, applying appreciative questions to mentor-mentee conversations as an innovative new practice of AI aimed to bring a "positive, democratic and entrepreneurial substance" to the hotel's existing mentoring relationships and enhance its employee engagement for sustainability. We found AI to be a useful theoretical perspective for this initial exploration of an area that was little researched or theorized.

3 Methodology

3.1 Theoretical Context

A research design that integrates a theoretical perspective, an advocacy or ideology, has a transformational value or action-oriented dimension and is called a transformative design (Greene and Caracelli 1997; Creswell el al. 2003). In this research, a transformative theoretical perspective was used within a sequential transformative mixed method design in a participatory AR study. We preferred the sequential transformative model described earlier due to its use of distinct phases that facilitate the implementation, description, and sharing of results. Yet, this required time to complete different data collection stages, as a sequential transformative strategy is a two-phase project with a theoretical lens overlaying the sequential procedures (Creswell 2009).

Our research consisted of generating entrepreneurial, unsupervised, individual, group and organizational action toward a more engaged workforce of the hotel. The data collection decisions required an awareness of the cultural values and practices of employees related to sustainability in the specific population of hotel employees. AR was chosen for this study due to its participatory and collaborative nature and the fact that "the subjective views of participants are of most importance in enacting change" (Ivankova 2015, p. 286). AR is viewed as a combination of empirical knowledge (knowledge derived from experience) and rational procedures (knowledge derived from scientific reasoning) that require multiple sources of evidence (Christ 2010). Thus, the research design chosen for our research exists within the post-positivist and constructivist paradigms as one phase of the research connects and informs later phases (Creswell and Plano Clark 2011).

A mixed methods design was appropriate, as both quantitative and qualitative data would help to provide the best understanding for our research questions, and together with our desk research, benefit from triangulation of the findings, adding scope and breadth to our study. Mixed methods research can be defined as the collection, analysis, and integration of quantitative and qualitative data in a single study or in a program of inquiry (Creswell and Plano Clark 2007). Mixed methods research is frequently cited as lending itself to a transformative framework (Mertens 2003), or applying an advocacy lens (Creswell 2003). Yet, contributions that address mixed methods designs in AR are still scarce (Mertens et al. 2010; Ivankova 2014), and do not systematize how qualitative and quantitative methods are integrated with participatory dynamics of AR.

3.2 Research Design and Scope of the Study

Our research was carried out in two phases between April 2014 and May 2016. It had an initial quantitative phase that relied on primary data, which preceded the participatory action phase, with the aim of helping its design. The objective of this first phase was to describe the hotel's sustainability culture in terms of the differences in employees' perceptions of a corporate culture of sustainability across the ranks of the hotel. Differences in perceptions were sought on certain cultural elements that related to leadership commitment, organizational systems, organizational climate, change readiness, and external and internal stakeholder (employee) engagement. Measuring these aspects of corporate culture was essential as we cannot manage or transform what we cannot measure.

This first phase was followed by a second qualitative and participatory AR phase that built on the earlier phase. Our transformative theoretical perspective of AI and the ideology of empowerment shaped our research questions and aimed at bringing voices of various frontline/operational employees—a group of mentors—to senior management's attention, co-creating knowledge through action and reflection. This was essential to better understand employee engagement as a process that is changing as a result of interacting with hotel employees as co-researchers. The focus of this second phase was on the author's experience of engaging with the hotel's senior management and mentors as critical practitioners in the AI process.

Figure 2 is a visual diagram that was developed to capture the flow of the research activities using the sequential transformative mixed methods methodological framework for AR. As the figure depicts, our design followed a schema

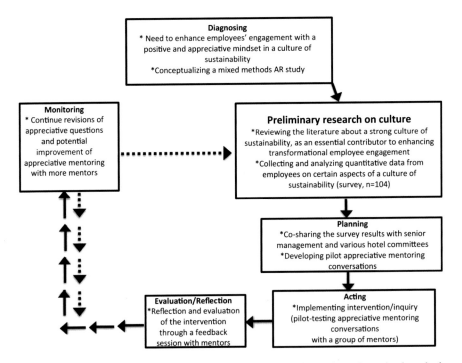

Fig. 2 The flow of our research activities using the sequential transformative mixed methods methodological framework for action research

that we named QUANTITATIVE-QUALITATIVE PARTICIPATORY, and thus shows how the two methods can be sequentially integrated within an AR project. Practical participation in this research is characterized by the involvement of participants such as senior management and employees. While solid arrows indicate the sequence of the phases in the research process, dashed arrows show other possible iterations of the research activities within an AR cycle.

During the Diagnosing phase, when the issue or organizational goal of enhancing employee engagement was identified in the hotel, mixed methods served to conceptualize the issue and identified the rationale for investigating it by using both quantitative and qualitative methods. In order to develop a plan of action/intervention, the researcher conducted a preliminary quantitative assessment of the corporate culture of sustainability using a survey method, which, in Kurt Lewin's (1948) terminology, could be called a fact-finding phase. A systematic and integrative collection and analysis of quantitative data on cultural attributes such as leadership commitment, organizational climate, organizational systems, change readiness, external stakeholder engagement and employee engagement during this phase helped generate thorough interpretations of the assessment results.

During the Planning phase, the action objective or expected outcome of enhancing employee engagement was set and an action/intervention of appreciative mentoring conversations was designed together with the HR team, having received senior management's support for the intervention. During the Acting phase, we implemented the action/intervention plan of pilot-testing appreciative mentoring conversation.

Then it was necessary to conduct a rigorous evaluation of the action/intervention to see the range of learnings of mentors through the process of briefing, guiding feedback, self-reflection and self-assessment of their pilot mentoring conversations. The use of mixed methods during the Evaluation phase of the AI cycle involved collection of qualitative data in the form of mentors' opinions in the mentoring conversations and interpretation of these qualitative results. The AI cycle was an improvisational process, as it progressed as a natural, bottom-up initiative of mentors rather than a top-down senior management decision.

Based on the new set of ideas, images and collective aspirations that were generated during the mentors' action/intervention evaluation, the group of mentors, the HR manager and the researcher made a decision to further revise or test the action/intervention plan of pilot appreciative mentoring conversations in the prospective Monitoring phase. The decision was to continue with the improvisational intervention and subsequently conduct more mixed methods evaluation of the intervention outcomes, which could lead to further refinement of the action/intervention plan of enhancing employee engagement throughout the hotel. The alternative was to return to the Preliminary research/Fact-finding phase and conduct more needs assessment on corporate culture and change the action plan based on the new mixed method inferences. However, in view of the limited time frame of our study, we had to reject this alternative.

3.3 Research Site and Population

Our research site was a mid-century central London hotel in the luxury hotel market recognized for its sustainability initiatives and commitment to employee-driven committees and taskforces that drive its employee engagement. Wholly owned by an overseas entrepreneur, the luxury hotel accommodates 416 guestrooms situated across 18 floors as well as one of Europe's largest and most flexible meeting and banqueting facilities. The luxury London hotel is also officially rated as one of the AA[2]'s most sustainable hotels in London and a *Times Top 100 Best Companies to Work for* company, with more than 450 employees. During the research period, the hotel was undergoing a huge renovation of its interiors and the outside facade, which aimed at making the hotel even more sustainable by 2017.

The hotel has a vision statement, "We always care", which emerges from a more holistic societal concern than the industry's traditional focus on guests. Its primary sustainability focus is on delivering educational hospitality (Hashmi and Muff 2015), which involves engaging guests in eco-friendly or sustainable practices, engaging employees in sustainability initiatives as well as engaging with external stakeholders in the form of sectorial/cross-sectorial collaborations that focus on the industry's need for skilled labor. The hotel aspires to be the most engaged hotel ever in London by 2017 when the hotel will turn 50. The hotel routinely got its employee engagement measured by Times Top 100 and several other in-house engagement surveys every year.

3.4 The SCALA Survey

To understand the differences in employees' perceptions of an organizational culture of sustainability across the ranks of the hotel, we used a survey method. We preferred the survey method to interviews, as our purpose was to generalize from a sample to a population so that inferences could be made about specific cultural characteristics, attitudes and behavior of the population (Babbie 1990). Descriptive data, behavioral data, and opinions, which are the keys to determining detailed and personal information about large populations, are obtainable with a known level of accuracy, and defined and determinable reliability only through the survey research process (Rea and Parker 2005). Thus, in this research, a first quantitative phase relied on primary data of a Sustainability Culture and Leadership Assessment

---

[2]The Automobile Association (AA) is the only pan-Britain assessing organization and is the British Hospitality Association's Patron Supplier for quality rating and assessment to the hospitality industry. In collaboration with Visit England, Visit Scotland and Visit Wales, the AA has developed Common Quality Standards for inspecting and rating hotels and guest accommodation through the AA accommodation scheme.

(SCALA) survey, which preceded a participatory/AR phase, with the aim of justifying the need for an AR intervention and contributing to its design.

The SCALA survey was developed based on a broad range of research methodologies including case studies, in-depth field research, survey research, archival research and empirical analysis. It is based on research that was conducted throughout 2010 and 2011, in which more than 200 interviews in more than 60 companies were administered to explore how sustainable companies were innovating for the development and execution of sustainability strategies. The SCALA was appropriate for our own research as it served to characterize the hotel's sustainability culture and leadership profile, which laid the foundation for a potential AR intervention on organizational transformation.

In the first phase (QUAN), we predicted a 20% response rate when figuring sample size. The survey was conducted in May 2014 over a 5-week period, and was sent to all employees in the hotel. The response rate was 22%, with 104 employees having responded out of an employee population of 465 at the time. Of the 104 survey responses, 13 were completed online and 91 were hard copies. The survey was administered online for the management team and as hard copies for supervisors and frontline staff in consideration of their computer access limitations. The SCALA survey instrument was composed of items pertaining to culture and leadership related to sustainability, and consisted of three sections and a total of 43 questions. While the questions in Section I were general questions on location, gender, age group and organizational rank of the employees, questions in Section II were Likert-type[3] statements about cultural elements of sustainability, and the rest of the questions in Section III were semi-open and open-ended questions related to the business sustainability positioning of the hotel. Section II questions focused on the following six categories of a culture of sustainability: organizational leadership, organizational systems, organizational climate, change readiness, engagement with internal stakeholders and engagement with external stakeholders.

This quantitative phase allowed us to understand and describe the culture of sustainability of the hotel in a short period without interrupting the hotel's schedule. The statistical analysis was carried out using SPSS—a software package used for statistical analysis, mainly in social science. The findings served to characterize the hotel's sustainability culture and leadership profile, and were also analyzed against differing functional levels. Understanding a culture of sustainability by position level was essential; as previous research shows that those at the highest levels in the organization have the most positive impressions of their companies' CSR or sustainability initiatives (Stawiski et al. 2010). The three different position levels identified were: executive/senior managers, middle managers (department heads) and frontline/operational staff (including supervisors and team leaders).

---

[3]A Likert-type statement is one that the respondent is asked to evaluate by giving it a quantitative value on any kind of subjective or objective dimension, with level of agreement/disagreement being the dimension most commonly used.

3.5   The Mentoring for Hearts and Minds (MHM) Project

During a presentation of the SCALA survey results to senior management and various committees in the hotel such as the Management Action Group (MAG) and Internal Communication Committee (ICC), a suggestion emerged to have a follow-up project: the 'Mentoring for Hearts & Minds (MHM)' project. The aim of the MHM project was to pilot-test appreciative interviews with a small group of pioneering mentors in their monthly mentoring conversations with their mentees, to enhance employee engagement for sustainability. This paper focuses on only the first round of the appreciative mentoring conversations of the MHM project, which formed the foundation block for a series of appreciative mentoring conversations that the hotel subsequently engaged in. To understand the transformation towards a stronger culture of sustainability and enhance employee engagement, we opted for an AR method. We were interested in "hearing voices" of participants, engaging them in the generation of knowledge through the iterative process of action and learning, facilitating their ability to develop successful action strategies, and linking the research results to practical ends (Hammersley 2000).

The MHM project set sail with a pioneering mentor group—an AI Learning Team—embracing the topic of enhancing employee engagement. The mentors played an important role in this research by serving as change catalysts towards enhanced employee engagement. The AI Learning Team of mentors consisted of the human resources manager, the head chef, a food and beverage (F&B) cost controller, an F&B outlet supervisor and a senior guest-relations officer. The team was intentionally representative of the "whole system" of mentors so that different perspectives and living knowledge of participant mentors would be included in testing the AI process with the mentees. The small team of mentors and mentees represented a larger group of frontline employees that needed to be engaged through informal monthly conversations.

The introductory meeting (briefing)

The team's first task was to convene for an hour and a half to frame what they most wished for in terms of engaging with each other, discuss the type of potential appreciative questions to ask the mentees, and agree upon the overall AI process to follow. The team agreed that the hotel's existent monthly mentor-mentee conversations might serve as a useful platform for positive transformation towards a more positive and engaged hotel culture. A review of an interview guide, which included a list of potential appreciative questions related to employee engagement, was carried out and the team discussed each question to clarify understanding. Finally, questions and comments regarding the AI process were shared and a date for a feedback session was scheduled.

Conducting appreciative interviews

The next step was to undergo inquiry time in the field with mentees. Over 5 weeks, all five mentors held monthly conversations with their mentees, engaging in appreciative interviews with the help of the agreed appreciative interview guide. In any AI process, it is important to have an appreciative interview guide for a more effective and efficient appreciative interview. Appreciative interviews mostly focus on discovery questions to search for, highlight and illuminate those factors that give life to the organization, yet in these interviews there were also questions related to the *Dream*, *Design* and *Destiny* phases of the 4D AI process. Giving mentors the opportunity to choose from a mixed pool of appreciative interview questions relating to different phases of the AI process helped them to improvise different types of questions involved in each of the 4D's of the AI process at their natural pace. Indeed, each AI process is unique based on its context, purpose, the type of people involved, and the skill and preferences of those leading or facilitating (Cooperrider and Srivastva 1987).

Feedback meeting (debriefing)

The feedback session enabled the mentors to share as a team, the key themes that had emerged during the appreciative interviews, and reach a consensus on the next course of action to sustain the momentum with appreciative mentoring conversations. After a quick check-in from the participants, each mentor elaborated on his or her appreciative interview experience with his or her mentee focusing on their feelings and learnings involved in the process, and then on the key insights they had gained during the appreciative interviews. Participants in the exercise stated that the main learning from the process was not so much the knowledge generated about employee engagement, but rather observations and feelings experienced during the appreciative interview process. The meeting concluded with consensus on action planning of a reworded appreciative interview guide and an AI taskforce that would champion employee engagement throughout the hotel, which the mentors considered essential in order for the appreciative mentoring conversations to be effective and further be embedded in the hotel's culture.

4 Findings

4.1 The SCALA Survey Findings

The SCALA survey highlighted that the London hotel's sustainability goals; initiatives and achievements appeared to have been generated from the organizational values, which almost every staff member truly believed in. Such a value-driven sustainability orientation justified the hotel's strong culture of sustainability. However, it was interesting to note that there were noteworthy differences in perceptions of positivity from the top of the organization down through the ranks of the hotel. The survey results clearly showed that, in line with our expectations, executive/senior managers were more positive than middle managers, who in turn, were more positive than frontline staff on almost all of the interval scale items relating to organizational leadership, organizational systems, organizational climate, change readiness, internal stakeholders and external stakeholders. Level of positivity was associated with the degree of agreement to the statements in the Likert-scale questions. The results demonstrated higher self-esteem, confidence and commitment on the management side. They further showed that top-level managers had the strongest sense of ownership of sustainability initiatives and commitment to sustainability as decision-makers involved in the hotel's critical sustainability decisions.

The most striking item in the SCALA results was that an average of 67% of respondents believed that the hotel had a clear strategy for engaging all internal stakeholders (employees) in its sustainability efforts, although middle managers (84%) were significantly more positive than frontline staff (57%, $p < 0.02$). This difference in positivity, especially between the middle managers/supervisors and frontline staff, did indicate a potential opportunity to the hotel's senior management to bridge the gap in positivity throughout the levels and further enhance employee engagement—the hotel's overarching HR goal—down to its line functions.

4.2   The Mentoring for Hearts and Minds (MHM) Project Findings

In the MHM project, appreciative interview questions related to employee engagement were integrated into the hotel's mentor-mentee conversations to bring a "positive and democratic stance" to the hotel's existing mentoring relationships, and thus enhance employee engagement for sustainability. The management team and the co-researcher believed that these informal mentor-mentee conversations carried huge potential in improving the current manager-subordinate relations and engagement in the hotel. The appreciative conversations between mentors and mentees not only engaged them in a spirited conversation about the hotel's greatest potential and strengths relating to employee engagement, but also about the strategic future of the hotel, namely, their images of an ideally engaged hotel and their statements of how they plan to organize themselves in pursuit of their dreams.

While the original intent of the appreciative questions was to delve into as many personal stories of the mentee based on their perception of employee engagement, one of the important insights by all the mentors was that mentoring experience and the extent of relationship with mentee played a crucial role in determining the depth of their conversations. Related to this finding, all mentors also mentioned the importance of the role of communication in building a healthy relationship with their mentee. The sharing of expectations and learning between the mentor and mentee were noted as a way to shape the relationship. Another insight from the appreciative interviews was that, where new mentees and mentors with little experience were concerned, it would be too soon to try out *all* the appreciative questions of the appreciative interview guide in one go, within the average mentoring time of one to one and a half hours. Last but not least, mentors mentioned that in the interview process, they liked and enjoyed *Discovery* and *Dream* questions more than *Design* and *Destiny* questions, since they thought these questions provided mentors with positive and more interesting information about the mentee.

The feedback session also led to some immediate outcomes in the form of innovative action. During the post-feedback period, the AI Learning Team worked together and reworded the appreciative questions in simplified language, to make them user-friendlier. Appendix 1 shows the initial questions versus reworded questions, to better see the changes made. It was also decided in the feedback session that the revised appreciative interview guide was to be tested in the following round of monthly mentor-mentee conversations. Yet, the consensus was that it would be used in mentoring conversations in a flexible way as before, namely, giving mentors the flexibility and discretion to choose a certain number of questions from each set of questions relating to different AI phase. This was deemed essential in consideration of the mentor-mentee relationship, differing mentor-mentee pair conversational needs, and most importantly, the surprises and improvisations that would occur on the way. Giving mentors the option to choose their own questions from a predetermined set of appreciative questions did not pose any issues on comparability of data as long as mentors used questions from each section of the interview guide. Each appreciative mentoring conversation was regarded as a unique, customized case to be reflected upon subjectively.

It was also decided that the AI Learning Team should meet and review feedback on the reworded questions with the HR team, and further share the tool with other members of senior management. This was to enable senior managers to potentially use appreciative questions in any prospective manager-subordinate performance review conversations also. As such, a taskforce consisting of mentors and the HR team was set up to drive employee engagement further up through the management layers of the hotel.

5 Discussion

Sharing SCALA findings both in senior management meetings and the hotel's various team meetings resulted in senior management's willingness to pilot-test appreciative interviews for more engaging and fulfilling mentor-mentee conversations at the hotel. This new idea of integrating appreciative questions to mentor-mentee conversations can be considered an innovation of this hotel, which, through managerial group reflection on the hotel's sustainability culture, emerged as a collaborative decision, rather than a top-down managerial decision solely owned by the general manager of the hotel. Indeed, in every culture or organization, there are marginalized voices and these voices are often the ones where important innovations reside (Whitney and Trosten-Bloom 2003).

We might argue that the improvisational approach of the appreciative interview process appeared to disclose new ideas, images and collective aspirations of mentors; altered their social construction of employee engagement that was initially limited to employee engagement surveys only; and, in the process, made available decisions and entrepreneurial actions that were not available or did not occur to them before. This shows that innovative, generative knowledge can be created through empirical work that takes a constructionist, reflexive stance (Schall et al. 2004). The innovation and improvisation of integrating appreciative questions into the hotel's monthly mentoring one-on-one conversations required some form of infrastructure in the form of cross-team networking and entrepreneurial capability among the mentors, which created empowerment and commitment on the mentors' side, and further enhanced the mentors' interactions with senior management increasing employee engagement. This supports Weick and Westley's (1996) argument that change outcomes are not created quickly, but incrementally in small wins that can be built on to transform large systems over time.

The MHM project demonstrated how an organization, at the intersection of innovation, transition and sustainability, found its positive core—the power of their mentoring one-on-one conversations—through a generative appreciative interview process applied as an iterative entrepreneurial micro-practice in mentoring conversations. The revised appreciative interview guide and the AI taskforce emerged without any formal mechanism imposed from senior management. Indeed, evidence suggests that change cannot be imposed or controlled from the top or from the outside (Senge 1990; Sullivan 2004). As our research indicates, improvisation can potentially become a key element of innovation, entrepreneurial capability and transformational change at the grassroots level in any organization. Yet, this also raises the question whether AI can successfully be realized with participants who already possess a high entrepreneurial and learning capacity. If so, should AI be limited to such target groups, or should one settle for mediocrity when such capacity is not pre-existent?

The outcome of the pilot appreciative mentoring conversations highlighted the fact that mentors played a critical role in mediating the micro-and macro-power dynamics within the hotel. More specifically, the subjective interpretations and perspectives of mentors conducting the pilot appreciative conversations offered a learning platform to help envision a collective structural decision at the macro organizational level to apply appreciative interviews to other conversations throughout the entire hotel. Whitney and Trosten-Bloom (2003, p. 217) describe this systemic application of AI to conversations, programs, processes, and systems throughout the entire organization as the third dimension that enhances the organization's capacity for ongoing positive change; the first dimension having to do with recognition of what has been learned and transformed in the process to date… and the second dimension being the initiation of cross-functional, cross-level projects and innovation teams… all of which launch a wide-range of action-oriented changes.

Beutel and Spooner-Lane (2009) asserted that the success of mentoring relationships lies in the skills and knowledge of the mentors; yet this also necessitates developing professional-personal relationships. Thus, for mentoring relationships to be successful and helpful to all three parties (mentor-mentee-organization), our research indicates that there is a need for a positive, appreciative and entrepreneurial approach to mentoring. An appreciative and guided approach to mentoring can assist mentors in their practices and help to build effective professional relationships. Although a mentor acts as an official of the organization and thus usually has some authority over the mentee, the appreciative paradigm is concerned primarily with the growth and development of the mentee, thus championing the relational dynamic between mentor and mentee. Interestingly, there is also a reciprocal process involved in the appreciative paradigm. In the most basic sense, by learning about the aspirations (dreams), challenges and experiences of the mentee, and collaborating with him/her in co-constructing his/her dreams, the mentor can be energized and advance his/her practice of mentoring to another level.

In consideration of these symbiotic benefits for a mentor-mentee pair and an organization, we propose that appreciative mentoring can be a powerful mechanism for mentors to intentionally reframe their interactions with mentees and senior management in a positive and energizing way.

6 Conclusion and Implications

While many companies are acknowledging the importance of sustainability, the implementation of an organizational sustainability strategic integration challenge such as employee engagement, remains a challenge for many businesses. As demonstrated by both the empirical and the AR evidence provided throughout the discussion, a culture of sustainability is conducive to effective implementing sustainability strategies, which require solid organizational transformation in the win-win-win outcomes for the environment, society and businesses. We suggest that achieving organizational change begins with leadership commitment of the management team, but these efforts must be complemented by improvisational and entrepreneurial operational practices such as mentoring one-on-ones that permeate the entire organization.

When enriched with appreciative questions, mentoring conversations can generate the positive energy that is needed to carry out organizational change. Because the changes proposed are based on the mentor-mentee pairs' own experience and relationship, and are developed by the stakeholders themselves, mentors and mentees as organizational members have a significant investment in the outcome. Also, because the AI process is cyclical, unlike a linear process that is carried out and completed, at its best, appreciative mentoring yields an "appreciative organizational culture" or an "appreciative mentor and mentee community" where many members develop their entrepreneurial skills at designing a future that carries forward the best of the past in an innovative way.

This research serves as an intentional application of mixed methods in action research and contributes to the existing pool of empirical mixed methods studies from various disciplines. It sets the stage for further transformative mixed methods studies to determine the impact of appreciative mentoring on the mentor-mentee relationship and to reinforce its applicability to different mentoring programs in different contexts. In this sense, we hoped to stimulate the use of a transformative perspective in research and ultimately help address a pressing organizational sustainability strategic integration challenge such as employee engagement, prevalent in our society today. Our research also demonstrated the importance of integrating a quantitative method to AR, which has long been ignored in the field (Chandler and Torbert 2003, p. 148). In this respect, we hope to have contributed to enriching AR literature by showing how a quantitative method could be integrated with participatory dynamics in AR designs in a way that data did not merely reflect an external world independent from actors/researchers.

More empirical research is needed to understand how appreciative mentoring may impact an organization's engagement level over time through changes in mentors' and mentees' perceptions on different aspects, such as their mentoring abilities to demonstrate positive interpersonal skills, to share stories, dreams and expectations, and to co-construct and sustain common roles that both mentors and mentees can undertake. Finally, further research could also focus on comparing how different variations of AI are more or less effective in mentoring programs.

Appendix 1

Initial Versus Reworded Appreciative Interview Guide

Initial Appreciative Questions	Reworded Appreciative Questions
Introduction (Rapport-building)	**Introduction (Rapport-building)**
1. I'd love to hear your initial attraction to Lancaster London. What inspired you to apply?	1. What attracted you to Lancaster London and/or what inspired you to apply?
2. Please describe a high point experience—a time when you felt most alive, happy and engaged as a hotel employee.	2. Describe a good experience since you began working at Lancaster?
3. What do you most value about: Yourself? Your work? Your hotel of choice?	3. Describe a time when you felt happy and engaged at work?
4. What is it that you plan to do with your precious life, and how does this relate to Lancaster London as your workplace?	4. What do you value about your work and your choice of Hotel?
Discovery—"Best of what is and what has been"	5. What is your ambition in the long run?
1. Recall a time when you worked with someone you considered as an engaged employee/colleague/manager. Describe the situation. What made him/her "engaged" for you? Who was the employee and what did she/he do that made you feel he/she was engaged?	**Discovery: 'Best of what is and what has been'**
	1. Recall a time when you worked with someone that inspired you as an engaged employee
	2. What was the situation
	3. What was it that made you feel that they were committed or engaged?
	4. What do you most value and/or appreciate about employee engagement at the Lancaster?

(continued)

(continued)

Initial Appreciative Questions	Reworded Appreciative Questions
2. What do you most value and appreciate about employee engagement at the Lancaster London?	5. Tell me about a time when you were committed or engaged at work. How did you feel? How did your colleagues respond?
3. Tell me about a time when "you" were the source of engagement. What created the sense of engagement? How did you feel at the time? How did others respond?	**Dream—"What might be"**
Dream—"What might be"	1. How can/could your manager help you to realize your potential or reach your goals and aspirations?
1. What should be the ideal engagement for helping you realize potential and present your achievement at the Lancaster London?	2. How can/could the hotel help you to realize your potential and reach your goals and aspirations?
2. What could work well for employee engagement in the hotel's future endeavors? What might be?	3. What do you think the hotel could do to improve employee engagement in the future?
Design—"What should be"	**Design: 'What should be'**
1. What will make the Lancaster London the most engaging workplace?	1. What in your opinion should the Lancaster London do to make it a more engaging workplace?
2. What hotel initiatives and processes would we actually need? Which ones would benefit us the most?	2. What hotel initiatives and/or processes do you think we need and which ones would benefit us the most?
3. Who will be involved in making the hotel the most engaging workplace?	3. Who is responsible for making the hotel the most engaging workplace?
4. How can we make the Lancaster London the most engaging workplace?	4. What will you do to make it the most engaging workplace?
5. What will we do (rather than need or want) to make it the most engaging workplace?	**Destiny: 'What will be'**
Destiny—"What will be"	1. What can we do together to realize your potential?
1. What is your first next step to help make the Lancaster London the most engaging workplace you dream of? *(Employee is asked to write a personal note to herself/himself, after having reflected on the following questions:*	2. What help/resources do you or we need?
* *So what can we do together?*	3. Who do we need to involve?
* *What help/resources do we need?*	4. How will we know/measure if you have accomplished your goal?
* *What people need to be involved?*	5. What will change as a result of your/our next steps?
* *How do we know when we've accomplished my/our goal?*	6. What existing initiatives do you choose to pursue based on your motivations?
* *What will change as a result of our/my next steps?*	
* *What existing initiatives do we choose to pursue based on our/my preferences and motivations?*	

References

Armstrong M (2010) Armstrong's essential human resources management practice: a guide to people management. Kogan Page, London, UK

Babbie ER (1990) Survey research methods. Wadsworth, Belmont, CA

Baltas A (2013) Managing in Turkish culture: acquiring global success with local values. Remzi Kitabevi, Istanbul

Barrett FJ, Fry RE (2005) Appreciative inquiry: a positive approach to building cooperative capacity. Teos Publications, Chagrin Falls, OH

Beutel D, Spooner-Lane R (2009) Building mentoring capabilities in experienced teachers. Int J Learning 16(4):351–360

Bierema L (2000) Moving beyond performance paradigms in human resources development. In: Wilson AL, Hayes ER (eds) Handbook of adult and continuing education. Jossey-Bass, San Francisco, CA, pp 278–293

Brinson J, Kottler J (1993) Cross-cultural mentoring in counselor education: A strategy for retaining minority faculty. Counselor Educ Supervision 32:241–253

Bushe GR (2001) Five theories of change embedded in appreciative inquiry. In: Cooperrider DL, Sorenson P, Whitney D, Yeager T (eds) Appreciative inquiry: an emerging direction for organization development. Stipes, Champaign, IL, pp 117–127

Bushe, GR (2005) Re-visioning OD: a post-modern reconstruction. Conference on social construction: a celebration of collaborative practices, Taos Institute, Taos, NM, Oct. 6–9 2005

Bushe GR, Coetzer G (1995) Appreciative inquiry as a team development intervention: a controlled experiment. J Appl Behavioral Science 31(1):13–30

Bushe GR, Kassam AF (2005) When is appreciative inquiry transformational? J Appl Behavioral Sci 41(2):161–181

Cameron KS, Quinn RE (2006) Diagnosing and changing organizational culture: based on the competing values framework. Addison-Wesley, Reading MA

Chandler D, Torbert B (2003) Transforming inquiry and action interweaving: 27 flavors of action research. Action Res 1(2):133–152

Chartered Institute of Personnel and Development (CIPD) (2011) Resourcing and talent planning survey. http://www.cipd.co.uk/surveys. Accessed 14 Dec 2015

Chouinard Y, Ellison J, Ridgeway R (2011) The sustainable economy. Harvard Bus Rev 89 (10):52–62

Christ, TW (2010) Teaching mixed methods and action research: pedagogical, practical and evaluative considerations. In: Tashakkori A, Teddlie C (eds), Sage handbook of mixed methods in social and behavioral research, 2nd ed. Sage, Thousand Oaks, CA, pp 643–676

Coghlan D, Brannick T (2010) Doing action research in your own organization, 3rd edn. Sage, London

Collins J (2001) Good to great. Harper Collins Publishers Inc., NY, pp 124–129

Cooperrider DL, Fry R (2013) Mirror flourishing and the positive psychology of sustainability. J Corporate Citizenship 46

Cooperrider DL, Srivastva S (1987) Appreciative inquiry in organizational life. In: Pasmore WA, Woodman W (eds) Research in organizational change and development 1: 129–169

Cooperrider DL, Whitney D (1999) Appreciative inquiry. Berret-Koehler, San Francisco, p 11

Cooperrider DL, Whitney DK (2005) Appreciative inquiry: a positive revolution in change. Berrett-Koehler Publishers, San Francisco, CA

Cooperrider DL, Whitney DK, Stavros JM (2003) Appreciative inquiry handbook 1. Berrett-Koehler Publishers, San Francisco, CA

Cooperrider DL, Whitney DK, Stavros JM (2005) Appreciative inquiry handbook 1. Berrett-Koehler Publishers, San Francisco, CA

Cooperrider DL, Whitney DK, Stavros JM (2008) Appreciative inquiry handbook for leaders of change. Berrett-Koehler Publishers, San Francisco, CA

Cooperrider D, Godwin L (2012) Positive organizational development: innovation-inspired change in an economy and ecology of strengths. In: Cameron KS, Spreitzer G (eds) The Oxford handbook of positive organizational scholarship. Oxford University Press, Oxford, pp 737–751

Crane A (1995) Rhetoric and reality in the greening of organizational culture. Greener Manage Int 11(12):49–62

Creswell JW (2003) Research design: qualitative and quantitative, and mixed approaches. Sage, Thousand Oaks, CA

Creswell JW (2009) Research design: qualitative, quantitative, and mixed methods approaches, 3rd edn. Sage Publications Inc, Los Angeles

Creswell JW, Plano Clark VL (2007) Designing and conducting mixed methods research. Sage, Thousand Oaks, CA

Creswell JW, Plano Clark VL (2011) Designing and conducting mixed methods research, 2nd edn. Sage, Thousand Oaks, CA

Creswell JW, Plano Clark VL, Gutmann ML, Hanson WE (2003) Advanced mixed methods research designs. In: Tashakkori A, Teddlie C (eds) Handbook of Mixed Methods in Social and Behavioral Research. Sage Publications, Thousand Oaks, CA, pp 209–240

Darwin A (2000) Critical reflections on mentoring in work settings. Adult Educ Q 50(3):197–211

Eccles RG, Ioannou I, Serafeim G (2012) The impact of a corporate culture of sustainability on corporate behavior and performance. Working paper 12–35, May 9 2012, Harvard Business School, Boston

Fredrickson BL, Branigan C (2005) Positive emotions broaden the scope of attention and thought-action repertoires. Cogn Emot 19(3):313–332

Gichohi PM (2014) The role of employee engagement in revitalizing creativity and innovation at the workplace: a survey of selected libraries in Meru coutry - Kenya. Lib Philos Prac 1:1–33

Glisson C (2007) Assessing and changing organizational culture and climate for effective services. Res Social Work Practice 17:736–747

Greene JC, Caracelli VJ (1997) Advances in mixed-method evaluation: the challenges and benefits of integrating diverse paradigms (New directions for evaluation, n 74). Jossey-Bass, San Francisco

Hammersley M (2000) Taking sides in social research. Routledge, London

Hansman CA (2000) Formal mentoring programs. In: Wilson AL, Hayes ER (eds) Handbook of adult and continuing education. Jossey-Bass, San Francisco, pp 493–507

Hansman CA (2002) Critical perspectives on mentoring: trends and issues. (Information series no. 388) ERIC Clearinghouse on adult, career, and vocational education, Columbus, Ohio

Harter JK, Schmidt FL, Hayes TL (2002) Business-unit level relationship between employee satisfaction, employee engagement, and business outcomes: a meta- analysis. J Appl Psychol 87:268–279

Hashmi ZG, Muff K (2015) Evolving towards truly sustainable hotels through a "well-being" lens: the S-WELL sustainability grid. In: Gardetti MA, Torres AL (eds) Sustainability in hospitality: how innovative hotels are transforming the industry. Greenleaf Publishing, Sheffield, pp 117–135

Higgins MC, Kram KE (2001) Re-conceptualizing mentoring at work: a developmental network perspective. Acad Manag Rev 26(2):264–288

Ivankova NV (2014) Mixed methods applications in action research: from methods to community action. Sage, Thousand Oaks, CA

Ivankova NV (2015) Mixed methods applications in action research: from methods to community action. CA, Sage, Los Angeles, p 286

Jarnagin C, Slocum JW Jr (2007) Creating corporate cultures through mythopoeic leadership. Org Dyn 36(3):288–302

Johnson-Bailey J, Cervero RM (2001) A critical review of the U.S. literature on race and adult education: implications for widening access. J Adult Continuing Educ 7(1): 3–44

Johnson-Bailey J, Cervero RM (2002) Cross-cultural mentoring as a context for learning. New Directions for Adult & Continuing Educ 96:15–26

Kessler EH (2013) Encyclopedia of management theory: the appreciative inquiry model. Sage Publications, Thousand Oaks, CA. doi: 10.4135/978145227

Kristof-Brown A, Zimmerman R, Johnson E (2005) Consequences of individuals' fit at work: a meta-analysis of person-job, person-organization, person-group, and person-supervisor fit. Pers Psychol 58(2):281–342

Lewin K (1948) Resolving social conflicts: selected papers on group dynamics. Lewin GW (eds). Harper and Row, New York

Lewis R, Donaldson-Feilder E, Tharani T (2011) Management competencies for enhancing employee engagement. Chartered institute of personnel and development. London, UK. http://www.cipd.co.uk/hr-resources/research/management-competencies-for-engagement.aspx. Accessed 08 Oct 2015

Ludema JD, Fry RE (2008) The practice of appreciative inquiry. In: Reason P, Bradbury H (eds) The SAGE handbook of action research. SAGE Publications Ltd, London, pp 280–295

Mckee R (2003) Storytelling that moves people. Harvard Bus Rev 81:51–55

Mertens D (2003) Mixed methods and the politics of social research: the transformative-emancipatory perspective. In: Tashakorri A, Teddlie C (eds) Handbook of mixed methods in social and behavioral research. Sage, London

Mertens DM, Bledsoe K, Sullivan M, Wilson A (2010) Utilization of mixed methods for transformative purposes. In: Teddlie C, Tashakorri A (eds) Handbookd of mixed methods research (2nd ed), Sage, Thousand Oaks, CA

Miller-Perkins K (2011) Sustainability culture: culture and leadership assessment. Miller Consultants Inc. http://www.millerconsultants.com/sustainability.php. Accessed 22 Feb 2016

Miner AS, Bassoff P, Moorman C (2001) Organizational improvisation and learning: a field study. Adm Sci Q 46(2):304–337

Naisbitt J (2001) High tech-high touch: technology and our accelerated search for meaning. Nicholas Braely Publishing, London

Network for Business Sustainability (NBS) (2012) Innovating for sustainability: a guide for executives. http://www.nbs.net. Accessed 28 Feb 2016

Nielsen K, Randall R (2012) The importance of employee participation and perceptions of changes in procedures in a team working intervention. Work Stress 26:91–111. doi:10.1080/02678373.2012.682721

Ragins BR (1997) Diversified mentoring relationships in organizations: a power perspective. Acad Manag Rev 22(2):482–521

Rea LM, Parker RA (2005) Designing and conducting survey research: a comprehensive guide. Jossey-Bass, San Francisco, CA

Schall E, Sonia O, Bethany G, Dodge J (2004) Appreciative narratives as leadership research: matching method to lens. In: Cooperrider DL, Avital M (eds) Constructive discourse and human organization: advances in appreciative inquiry. Elsevier, New York, pp 147–170

Schaufeli WB, Bakker AB (2010) Defining and measuring work engagement: bringing clarity to the concept. In: Bakker AB, Leiter MP (eds) Work engagement: a handbook of essential theory and research. Psychology Press, New York, pp 10–24

Sekerka LE, Brumbaugh A, Rosa J, Cooperrider D (2006) Comparing appreciative inquiry to a diagnostic technique in organizational change: the moderating effects of gender. Int J Org Theory Behav 9(4):449–489

Sekerka LE, Zolin R, Smith GJ (2009) Careful what you ask for: how inquiry strategy influences readiness mode. Organization Manage J 6:106–122

Senge P (1990) The fifth discipline: the art and practice of organizational learning, revised edn. Random House, Milsons Point, Australia

Stawiski S, Deal JJ, Gentry W (2010) Employee perceptions of corporate social responsibility: the implications for your organization. Quick view leadership series. Center for Creative Leadership, USA

Sullivan E (2004) Becoming influential: a guide for nurses. Pearson/Prentice-Hall, London

Sundaray BK (2011) Employee engagement: a driver of organizational effectiveness. Eur J Bus Manag 3(8). Available at: www.iiste.org

Thatchenkery T, Metzker C (2006) Appreciative intelligence: seeing the mighty oak in the acorn. Berrett-Koehler, San Francisco, CA

Thomas DA (2001) The truth about mentoring minorities: race matters. Harvard Bus Rev 79 (4):99–107

Unsworth KL (2003) Engagement in employee innovation: a grounded theory investigation. Available at: http://eprints.qut.edu.au/3033/

Weick KE, Westley F (1996) Organizational learning: affirming an oxymoron. In: Clegg SR, Hardy C, Nord WR (eds) Handbook of organization studies. Sage, London

Whitney D, Trosten-Bloom A (2003) The power of appreciative inquiry: a practical guide to positive change. Berrett-Koehler Publishers Inc, San Francisco, CA, p 161

Woodman RW, Sawyer JE, Griffin RW (1993) Toward a theory of organizational creativity. Acad Manag Rev 18(2):293–321

Zahra SA (2011) Entrepreneurial capability: opportunity pursuit and game changing. 3rd Annual conference of the academy of entrepreneurship and innovation, June 15–17, p 1–39

Entrepreneurship, Innovation and Luxury: The Case of ANTHYIA

Miguel Angel Gardetti

Abstract Sustainable luxury not only requires social and environmental performance of real excellence throughout the value chain in organisations but also, local cultural contents such as the craftsmanship and local innovation. Today brands that will be considered luxury labels in the future are being created through innovation and entrepreneurship. More than simply observing reduced negative impacts of operations, entrepreneurs and innovators seek to solve social and environmental problems through the development or acquisition of new capabilities that are directly related to the challenge of sustainability. Anthyia is an organisation created by Ying Luo and Nicolai Nielsen. This company is a pioneer in the use of Ramie, a vegetable fibre from China. While it was founded in 2011 in the United States, it currently operates in Germany. This paper studies Anthya from the model of sustainable value creation developed by Professor Stuart L. Hart that integrates the environment, innovation, stakeholder management and potential for growth. Anthyia is focused on Ying's personal values and beliefs and on innovation. Ying and her partners develop the abilities that enable to position (and reposition) the firm for the future. Sustainable competences arising from ramie research is the key to the company's value creation. Anthyia is the combination of Ying's personal values and the Eastern perspective of ramie in a continuous creativity and innovation process. This is essential for the company's long-term existence and, since it is a dynamic concept, it requires managerial and organisational leadership that are typical of both Ying and Anthyia.

Keywords Luxury · Sustainable luxury · Anthyia · Entrepreneurship, innovation, Ramie, sustainable value creation

First and foremost, the author would like to thank **Ying Luo and Nicolai Nielsen**—Anthyia founders—for their valuable contribution to this case.

M. A. Gardetti (✉)
Centre for Studies on Sustainable Luxury, Av. San Isidro 4166, PB "A",
C1429ADP Buenos Aires, Argentina
e-mail: mag@lujosustentable.org
URL: http://www.lujosustentable.org

© Springer Nature Singapore Pte Ltd. 2018
M. A. Gardetti and S. S. Muthu (eds.), *Sustainable Luxury, Entrepreneurship, and Innovation*, Environmental Footprints and Eco-design of Products and Processes, https://doi.org/10.1007/978-981-10-6716-7_4

1   Luxury, Sustainability, Innovation and Entrepreneurship: An Introduction

Sustainable development is defined as a model whereby our present needs can be met without jeopardising the future generations' needs, as stated by the World Commission on Environment and Development (WCED 1987) report, *Our Common Future*. Sustainability implies, on the one hand, encompassing economic, environmental and social aspects—*which are dynamic and interact with one another, one receiving the influence of the remaining two*- and, on the other hand, *integrating the short term into the long term* with a multidisciplinary approach. This clearly conveys the systemic aspect of sustainability. As it can be noted, sustainable development uses a different approach to things. This is about a totally different notion of the world as presently seen, which includes justice, freedom and dignity (Ehrenfeld 1999). This implies acknowledging that sustainability is about well-being and long-term, and it is also about developing the "actual" awareness of its implications. This is a vision whereby we can create a way of "being" (Ehrenfeld 2002).[1] Other authors who have studied sustainable development at an individual level are Suzuki and Dressel (2002), and Cavagnaro and Curiel (2012). These authors basically explain that it is the evaluation of all human behaviours with a view to find those that contradict the development of a sustainable future, and add that no sustainable society is possible without sustainable individuals. That is, individual capacities seem to be at the heart of the issue.

Luxury is generally and historically associated with the idea of exaggeration, showing off, and a power position. It could be said that, while the first notions of possessing of goods as a status symbol began as early as the Middle Ages, it was not until the industrial revolution—in the late 18th century—that it gained the current significance (Maisonrouge 2013). Therefore, Featherstone (2014) argues that luxury has its own dynamism. As defined by Heine (2011), luxury is not a need and/or want but a desire that depends on different (cultural, economic and regional) contexts. This makes luxury to be an ambiguous concept (Low, undated). Since times immemorial luxury has been an example of opulence, power and social rank (Kapferer and Basten 2010). In "The idea of Luxury" authored by Christopher L. Berry in 1994 shows the luxury timeline and reflection of people's aspirations and society rules. As time went by the concept of luxury has changed at the same pace as the world has (Shamina 2011). Luxury is a question of how things are perceived and experienced, and why they have become "essential". However, according to Kapferer and Basten (2010), once people have tried luxury in any field, they find it hard to leave it. When purchasing power decreases, people start to cut expenses in any conventional product, but this is different in the case of luxury products. Along this same line, Featherstone (2014, p. 48) describes: "Luxuries are things which have power over us. They engage the senses and have the capacity to affect us by

---

[1]Ehrenfeld JR (2002) Sustainability by Choice, unpublished.

offering a range of pleasures. Yet luxuries also demand recognition, the capacity to know what they are, to tell the difference between luxuries and necessities."

While, according to Cloutier (2015), luxury conjures up visions of attractive, desirable lifestyle choices, it also represents a moral vice harmful to both the individual and the society. Thus the luxury/sustainability relationship consolidated in 2011 through a series of events and writings, some of which were academic (for example, Gardetti 2011; Gardetti and Torres 2013; Gardetti and Torres 2014a, b; Gardetti and Girón 2014; Gardetti and Muthu 2015; Gardetti and Girón 2016). This way—according to Kleanthous (2011)—luxury started to be a factor for people express their deepest values. In other words, sustainable luxury *is the concept of returning to the essence of luxury with its traditional focus on thoughtful purchasing and artisanal manufacturing, to the beauty of quality materials* (Gardetti and Torres 2013, p. 4). Based on this definition, luxury would be the driver of the respect for the environment and social development. This would include culture, art, local innovation, maintaining the legacy of local craftsmanship. Carcano (2013) along similar lines explains that "*luxury defends, among others, the artisanal level and natural materials*".

Thus, these definitions above clearly express the relationship between luxury and sustainability. And this is aligned with Wittign, Sommerrock, Beil and Albers statement (2014, p. 184): *high-end products are the antithesis of the modern world's throwaway culture*. And this relates to durability which is the core of sustainable development and luxury as well. Durability is the enemy of the mass fashion industry based on planned obsolescence. In contrast, luxury is the business of perdurable value (Kapferer 2015).

There are people with a deep look at the environmental and social problems and who are motivated to 'break' the rules and promote disruptive solutions to these problems and many of them are entrepreneurs. There are several authors who have documented the relationship between personal values and sustainability, entrepreneurship and innovation, for instance, Young and Tilley (2006), Choi and Gray (2008), and Dixon and Clifford (2007). Some projects have developed an inclusive supply chain with poor and vulnerable communities that respects local culture, developing simultaneously environmentally sustainable products. Innovation provides the means to be more than a change. This is the creation of a new reality (Rainey 2004). However, many times success is the result of trial and error, a result of searching, following the wrong path, making mistakes. And getting things wrong sometimes is a necessary part of innovation (Hoffman and Lecamp 2015).

Through innovation, they are leaders who inspire, intellectually stimulate and have a very important and deep consideration for people and the environment. And the relationship between leadership and inspiration is vital for the achievement of change (Lips-Wiersma and Morris 2011).

First of all, this paper presents the creating value model, and then introduces Anthyia, Inc. Subsequently, such undertaking is reviewed under the above mentioned model and the paper finalises with the conclusions.

2 Methodology

To develop this case, the author collected publicly disclosed background and information about the company. This source was supplemented by conducting interviews to the company's founders.

3 Creating Sustainable Value

Strategies and practices for value creation can be identified by addressing a sustainable world with a business mind-set. As explained by Porter and Kramer (2011), corporate purpose needs to be defined once again with the idea of going beyond profit and based upon value sharing. This way a connection between social and growing economic aspects is set (Porter and Kramer 2006; Gholami 2011). Based on interactions, a corporation needs to identify its stakeholders first and then to see how value will be created (Freeman 1984; Freeman et al. 2007). A *sustainable enterprise* has potential to link the private sector and development including the poor, cultural diversity and ecology preservation (Hart 2005, 2007).

3.1 Creating Value

Spatial and temporal variables are used to create short and long-term value.

Focused on the temporal variable a company must control its present business and be able to create its future technology and marketplaces. Reversely, the spatial variable speaks about taking care of the "in-house" corporate skills, technology and capabilities, without overlooking the "external" stakeholders (Hart 1997, 2005, 2007; Hart and Milstein 2003).

Figure 1 shows these two variables depicted in four distinctive value-creation dimensions (Hart and Milstein 2003) which should be managed efficiently and simultaneously.

3.2 Global Sustainability Drivers

See below the four sets of global sustainability drivers indicated by Hart (1997, 2005, 2007) and Hart and Milstein (2003).

Set 1 refers to increased industrialisation and related impacts. An efficient use of resources and no pollution are key to sustainable development.

Set 2 is about civil society stakeholders vis-à-vis high performance companies. Along this line, companies need to be open, accountable and informed to attain sustainable development goals.

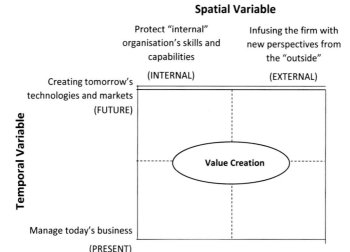

Fig. 1 Variables and Dimensions for Value Creation. *Source* Designed by the author (adapted from Stuart L. Hart and Mark Milstein, Creating Sustainable Value 2003)

Set 3 relates to emerging technologies and innovation resulting in radical and "disturbing" solutions and old technologies obsolescence.

And Set 4 is connected to a growing population. This is about the impact of globalisation in terms of local standalones, culture and environment and how this results in a slowdown in developing countries (Hart 2005, 2007; Hart and Milstein 2003). To meet sustainable development goals, the focus needs to be a long-run vision whereby socioeconomic and environmental matters intertwine.

3.3 The Sustainable Value Structure: Connecting Drivers with Strategies

A single corporate action is not enough to address such a multidimensional concept as global sustainability is. Any firm needs to handle the above mentioned four distinctive sets of drivers if it intends to create (sustainable) value (Hart 1997, 2005, 2007; Hart and Milstein 2003).

3.3.1 Growing Profits and Reducing Risks Through Pollution Prevention

Firms have an opportunity to cut down costs and risks and, at the same time, to develop eco-efficient abilities and prevent pollution (Hart 1995, 1997, 2005, 2007).

3.3.2 Enhancing Reputation and Legitimacy Through Product Stewardship

The stakeholders' point of view is embedded in product stewardship. Aligned with this, environmental and social impacts can be reduced all along the value chain, and companies can win by engaging stakeholders in their ongoing operations (Hart 1995, 1997, 2005 and 2007). According to Rainey (2004), this is a process of integration and opening which, in turn, builds confidence towards corporate sustainable development.

3.3.3 Market Innovation Through New Technologies

New sustainable technologies relate to innovations that go beyond present knowledge base (Hart and Milstein 1999). In the words of Quinn (1996), this is a process that puts an end to the past and cannot be reversed. As a result of this, companies can get to address social and environmental issues backed by their in-house development or otherwise companies can learn new specific capabilities (Hart 1997, 2005, 2007; Hart and Milstein 2003).

3.3.4 Crystallising the Growth Path Through the Sustainability Vision

The vision of being sustainable can create a fruitful road for future companies. And it provides the organisation members with the necessary guidance (Hart and Milstein 2003). This is what Rainey (2004) defined as moving from thought to action.

Sustainable development and short and long-term value creation are underscored in this model as shown in Fig. 2.

4   ANTHYIA Inc.

Anthyia was created to execute the founder's belief that ramie is the most sustainable luxury fibre for fashion and home textile industry. Anthyia's goal is to work with the right designers and the right factories who share the same belief in bringing the highest quality ramie yarn, fabrics and finished products to the world natural markets. With this goal in mind, Anthyia hopes to bridge the communication gap between the East and the West. Anthyia Inc. was founded in 2011 by Ying Luo, a Chinese born US citizen, currently a German resident. She calls herself "*a global resident*".

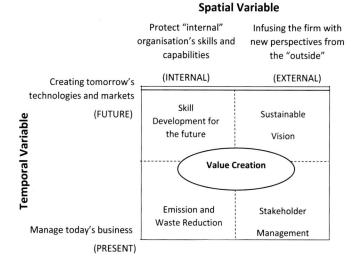

Fig. 2 Strategies and practices associated with value creation for short and long term and the essential elements for its development. *Source* Designed by the author (adapted from Stuart L. Hart and Mark Milstein, Creating Sustainable Value 2003)

4.1 The Founder and Her Values

In 1993, after graduating from Xiamen University, a distinguished university in China and with a track record of several years in international trade, imports and exports, Ying flew to the USA to pursue her MBA with a focus on international marketing. At that time, her friends and classmates back in China started to make fortunes by making and selling cheap products, trading stocks and spending fortunes on Western luxury brands, but she gave up her opportunities to make fortunes in China and went to the USA to learn about their "*big*" business ideas.

In 1996, on campus she was hired to work in a high tech company based in Silicon Valley. She lived in California both north and south for fourteen years, got married and had two children. During these 14 years she was well trained in cross-cultural business, quality production and international marketing. But she witnessed and got tired of the wasteful innovations of high tech gadgets that promoted overconsumption trends.

She saw first-hand this irresponsible manufacturing, low moral business competition, greedy materialism and earth killing pollution that the Western business had brought to China. People were demanding more and more and were wasteful and "*lost in their happy freedom dreams without thinking of the consequences of their behaviours*" (see Footnote 1).

She then felt an urgent need to join the wave of saving our children's future home and protecting our shared home earth.

When her two children became teenagers, with her husband, she decided to move back to China, so their children could learn their mother language. She also believed it was high time for her to start "*her own adventure*".

Because of her children, Ying discovered that most items are made in polyesters, and it is difficult to find natural fibre socks and bedding products. The idea was clear: she wanted to do something made of natural materials, but it was difficult in China.

Then, back to natural? China has never been so excited and diligent to copy Western luxury dream: skyscrapers can be erected overnight, and fastest trains and railways are making German engineering eclipsed. Big and small cities were nearly all surrounded by haze. The market is full of low quality products, so air, water and food are polluted. Nanchang is fortunately one of the cities in midland China, so it was not as polluted as the seashore cities.

Ying started to educate people: in her MBA class she often inserted green and sustainable life style trends in the USA, and she also wrote for the local lifestyle magazine introducing sustainability concept, she even participated in Jiangxi Education TV Station as a special guest for education and western lifestyle trend introduction. With this work, she was able to focus on environment and natural living topic that she likes.

4.2   Why Ramie?

Thanks to the convenient modern transportation built in China, it was easy for Ying to travel domestically. She travelled north and south, visiting the world biggest light industrial production centres in China. For example, Yiwu city produces three quarters of small commodities the world is buying; Keqiao textile centres holds, at least, one half of the low costs textiles for Chinese textile exporting, and Guangdong has 80% of accessories manufactured in China... This was so depressing for Ying. The products are so cheap that the cost control department in Yiwu management was really amazed. In Yiwu city, more than 2000 Western buyers are living and get ready for trading back and forth these cheap products to their countries. Ying was determined to show the world that China has more than cheap products. She set her eyes on ramie that she started to love, and be familiar with. She visited the fields in deep mountains, chatted with senior ramie engineers and farmers, and tested the existing fibres and products until she was totally convinced. So, in early 2011, Ying started visiting all the ramie fields in deep mountains, factories in Jiangxi, Hunan, Hubei, Anhui and Sichuan provinces, and Chinese main Research Centre of Bast Fibres (Institute of Bast Fibre Crops, Chinese Academy of Agricultural Sciences).

4.3   Ramie

According to Hamilton (2007) plant fibres can be divided into three categories depending on their source of origin: seed, bast, or leaf. As Ying said "*I was amazed by ramie which is totally ignored by the world*" (see Footnote 1). And, according to Hamilton (2007), this was due to the increased growth in the use of both cotton fibres (extracted from the seed) and synthetic fibres. Technically, the term "bast" refers to a complex tissue layer located under the bark, with multiple functions, such as fluid transport, nutrient storage, and structural support. The best-known bast fibre in the West, flax is an annual plant, as hemp and jute are. Ramie—*Boehmeria nivea*—is a perennial (Kadolph 2011) that sends out new stems each season and remains productive for several years.[2]

To many Chinese living along the Changjiang river, it is easy to see this wild plant growing everywhere all year long.[3] Changjiang river is the longest river in Asia and third longest river in the world. One hundred per cent of Changjiang river body is inside China, starting in Sichuan, going through Hubei, Hunan and Jiangxi and going out to the East Sea from Shanghai. Ramie plants are mainly growing in these four provinces due to their proper temperature and humidity (Kadolph 2011) and also in Anhui, Zhejiang and Taiwan (see Map 1). Photos 1, 2 and 3 show part of the ramie plant.

Please find below some significant facts about ramie:

– **Ramie is very sustainable plant.** Huge leaves purify the air and can be made into food or animal feeds, long stems provide fibres and wooden boards. Strong roots can be made into Chinese medicine and protect the soil from erosion. Ramie plants grow like grass, they can survive any shape of land: backyards, side roads, slopped fields. So, it is easy to cultivate. "*During original times, human didn't have anything to process fibres. Bast fibres are actually the only existing fibres that God created for us. Ramie is for Asian people. Once it is planted, it will continue coming back for 20 years*" (see Footnote 1). Ramie has 3–4 harvests per year, and most importantly, wild ramie grows mainly depending on rain water and as ramie is antibacterial there is no need to use pesticides.
– **Great fibre.** Ramie fibre is the strongest among all natural fibres (Kadolph 2011), and it is antibacterial. "*Even kings knew about this benefit and used these fibres to protect them after death*" (see Footnote 1). Scientists have found this in very ancient mummy pieces from noble people's and pharaohs' tombs. Even

---

[2]For Kozlowski et al. (2005a) Chinagrass, Ramie and Rhea are two plants. One of them, Chinagrass—Boehmeria nivea—raised confusion with the use of different terms, Chinagrass, Ramie and Rhea. This is the so called "white ramie". The other plant is likely a variety of the same species (Boehmeria nivea var. Tenacissima), though sometimes considered as a different species (B. Tenacissima). This is Ramie (Malay zanf) from Malaysia Island and Rhea from Assam.

[3]95% ramie grows in China due to its right temperature and humidity.

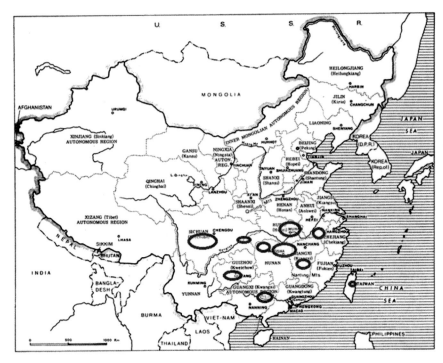

Map 1 Ramie growing areas in China. *Source* Prepared by the author. Based on a map prepared by the Food and Agriculture Organization of the United Nations—http://www.fao.org/docrep/005/AC862S/AC862S04.htm

Photo 1 Ramie field in China. *Source* Anthyia. Published with Anthyia's authorisation

Photo 2 Ramie stem. *Source* Anthyia. Published with Anthyia's authorisation

Photo 3 Ramie superior part. *Source* Anthyia. Published with Anthyia's authorisation

today in many Korean villages, they still use ramie fabrics to make funeral clothes. In China, people use ramie for summer sheets since it is cool at touch, absorbent and breathable, and fine ramie fibre has silk sheen and carries noble appearance (Aiguo, unknown year). And people—in China—still love cool ramie sheets, ramie mosquito nets, ramie clothing. Ramie fibre is still used for making grain bags, threads to make shoes, ship ropes and parachutes.

– **Crisis starts.** *"When the time for cheap and easily made plastic fibre arrived, bast fibres like ramie which needs lots of efforts were forgotten. Today more and more people clearly know that plastic fibres, heavy pesticides and water consumption cotton are not sustainable"* (see Footnote 1).

– **Ramie cultivation in China.**[4] There are two kinds of ramie cultivation in China: on steep slopes (depend on rain force), and on flat zones (depend on manmade canal irrigation). The quality of ramie fibre depends on the right ambient temperature and moisture. Although there are many ramie plants along the ChangJiang river, the best fibre comes from few villages. *"Anthyia only chooses fibre from DaZhou village* (see Map 2) *since the temperature and humidity helps the finest ramie to grow. Our ramie fibre counts can reach 2500 Nm[5] to 3000 Nm, while the average ramie fibre count from the remaining Chinese ramie fields is around 1200 to 1800 Nm"* (see Footnote 1). That is why you can see so many rough ramie fibres out there, but those ramie fibbers are not good for close to skin textiles. And that is also one reason why old ramie had a bad and cheap image in many people's mind. Photo 4 shows the lustre of the yarn used by Anthyia.

– **Ramie Harvesting and collecting**

In DaZhou county (see Map 2), *"Anthyia's partner signed up more than 2000 local farmer households, who plant their ramie in their fields or backyards, and harvest. They manually peel the stem skin, dry it in the air"* (see Footnote 1) (see Photos 5, 6, and 7), and then sell it to the Anthyia's partner. Most famers peel the skin manually, and in some larger area planting, famers can peel ramie skin with semi-automatic machines.

– **Ramie Degumming**[6]

The degumming process is very important to determine the softness of yarn and fabrics.

Since ancient times, human beings knew how to degum bast fibres in ponds. After many days and weeks, the soaked ramie skins would be rotten and then the rotten part would be washed away with water, and leave the strong fibres for drying. Alkali solutions and hours of cooking are used to increase the speed of taking the fruit gum away the fibres (Kozlowski et al. 2005a, b). The biggest task of this cooking degumming is to destroy the fibre resistance.

---

[4]Part of the data were surveyed by Kozlowski et al. (2005a, b).

[5]Nm is Number metric. It means the number of kilometres of yarn per kilogram of weight (Cohen and Johnson 2014).

[6]Part of the data were surveyed by Kozlowski et al. (2005a, b).

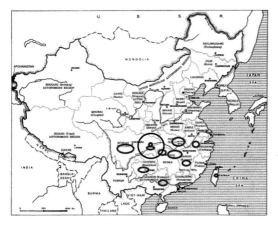

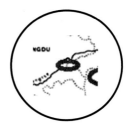

The small (blue) circle on the edge of the
red circle shows DaZhou Village

Map 2 DaZhou Village location. *Source* prepared by the author based on data from the Food and Agriculture Organization of the United Nations—http://www.fao.org/docrep/005/AC862S/AC862S04.htm, for the map, and from Aiguo, Zhu (w/o year) "Ramie (*Boehmeria nivea*) production and its diverse uses in China"

Photo 4 Ramie yarn stem. *Source* Anthyia. Published with Anthyia's authorisation

Most Anthyia fibres are degummed with 80–90% of biologic degumming (first soaked with micro bacteria for 7 h, and the fruit gum would have been eaten away 80–90%).

Photos 8 and 9 show preparations for degumming.

Photo 5 Peeling of Fibre from the Ramie stem. *Source* Anthyia. Published with Anthyia's authorisation

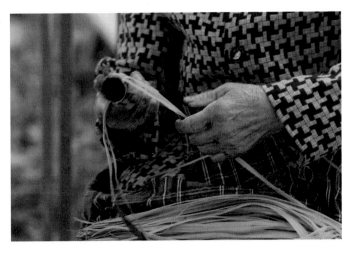

Photo 6 Peeling of Fibre from the Ramie stem (details). *Source* Anthyia. Published with Anthyia's authorisation

4.4 Creation of Anthyia

Anthyia was founded in the USA by the end of 2011. During her first four years studying and researching ramie in China, Ying hired Germany designers and pattern makers to set up a ramie home wear line, and attended organic fashion fairs in Germany twice a year to understand the end users in the leading global green textile

Photo 7 Drying the Ramie at air. *Source* Anthyia. Published with Anthyia's authorisation

Photo 8 Preparing for degumming. *Source* Anthyia. Published with Anthyia's authorisation

Photo 9 Cut off rough part of ramie before degumming. *Source* Anthyia. Published with
Anthyia's authorisation

market. There she discovered that Europeans were unaware of ramie and what they
were looking for.

Back to the ramie field, she got more familiar with the existing fibre and fabric
producing technology. In the meantime, she consulted senior ramie engineers and
research centres—the main Chinese Research Centre of Bast Fibres: Institute of
Bast Fibre Crops, Chinese Academy of Agricultural Sciences—regarding several
bottleneck challenges of current ramie fibre and fabrics: degumming, industrial
water pollution, fabric itching, easy to wrinkle and lack of resistance. These bot-
tlenecks plus the complex works and slow returns made people leave ramie fields.
Even farmers pulled out ramie plants and replaced them with popular and readily
sold vegetables and fruits. However, just at this time, Anthyia was founded.

4.4.1 Ramie Partner Searching

Before working with their current partners, Anthyia started researching biologic
degumming process in one factory suggested by the Bast Fibres Centre, which is
famous for non-chemical degumming. However their fabrics were too hard and
could not be made into high quality and comfortable close-to-skin items which is
really needed with chemical free degumming. From thousands of samples, Ying
chose natural dyed, softest woven fabrics. Then she made them into homewares and
showed them in Innatex and Green Show Room natural fashion shows in Germany.
After 3 years, Ying drew a conclusion, "*the fabrics were not good enough*" (see
Footnote 1). So she decided to go back to the yarn stage. After many years consulting research centres, factories and engineers, plus the experience gained at trade
shows, and self-testing of ramie clothes, Ying started to gain deeper understanding

Photo 10 Ying Luo with her partner Mr. Zhang Xiaozhu. *Source* Anthyia. Published with Anthyia's authorisation

of this fibre. She believed that ramie's potential "*was undiscovered*" (see Footnote 1). She flew to Japan where the biggest high-end ramie market was supposed to be. However, she did not find substantial differences. Therefore she decided to visit DaZhuo county, the place with the best environment for growing the highest counts of ramie fibres. When she met her actual partner Mr. Zhang Xiaozhu from Sichuan Jade ramie (see Photo 10), she found out that Zhang was also looking for a partner to help him get closer to the consumer market. This is how their cooperation started. Anthyia provided the standards of ecologic factory settings, natural and organic textile market requirements, and so they worked together to list what needed to be done.

4.4.2 Breakpoint Reaching and Products

From the end of 2011 to the end of 2014, Anthyia invested around 20,000 € per year which included research trips to bast fibre—not just ramie, but also hemp and flax (linen)—fields, factories, research centres, universities, individual professors, consultations; sample purchasing, sample making, designers and sewers' salaries, trades shows, international transportation, fashion/textile trade shows, production, promotion (free trials for designers, students, interviewers…), laboratories, etc. This investment did not include business costs, such as company set up, equipment, phone bills, home-office, gifts, materials, shipping fees.

On the other hand, Anthyia sales amounted to 26,000 € (8000 € in 2012; 15,000 € in 2013; and 3000 € in 2014). Anthyia shipping costs were too high so profit was small. Plus in 2014, Ying planned to move to Germany, so her business slowed down. Anthyia reviewed the ramie situation. There was an issue regarding GOTs

Photo 11 Ramie knitting fabric (made in Germany). *Source* Anthyia. Published with Anthyia's authorisation

certification. The shows Ying went to needed some certification, and she personally "*did not totally agree with this high cost certification*" (see Footnote 1). Therefore, Ying searched the right way to get a clean reputation. Even though she paid so much attention to our suppliers' environment and social responsibility, "*made in China was not a good term in Germany*" (see Footnote 1).[7]

By the time she moved to Germany, she completed educating the Germans about what ramie was all about. Then she decided to focus on improving knitting fabrics and understanding the market needs while she started sample production in Germany. Photos 11, 12, 13, 14, and 15 show some of Anthyia's products.

Ying started fabric trade shows in February 2016 and sales began in December 2016. Anthyia's sales reached about 50,000 € with about 40% profits mainly for Ramie yarn and fabrics. Despite this, "*Anthyia's R&D cost was high for new samples so net profit was nearly nil*" (see Footnote 1). The following year—2017— Anthyia expected more sales, and Ying expected to be breakeven in 2018/2019.

5 Creating Sustainable Value at Anthyia Inc. and Conclusions

To cut down risks Anthyia makes products in Germany, and has devised how to align its vision where value change and sustainability meets. Moreover, Anthyia included external stakeholders as the Research Centre of Bast Fibres (Institute of Bast Fibre Crops, Chinese Academy of Agricultural Sciences) and reached a differentiated position. Along with Mr. Zhang Xiaozhu, they created a commercial circle for ramie development providing artisan famers with their livelihood. Anthyia

---

[7]Anthyia had discussions with companies such as Hess-Natur (Germany).

Photo 12 Ramie tape (made in Italy). *Source* Anthyia. Published with Anthyia's authorisation

Photo 13 Ramie woman skirt. *Source* Anthyia. Published with Anthyia's authorisation

Photo 14 Ramie socks. *Source* Anthyia. Published with Anthyia's authorisation

Photo 15 Ramie buttons.
Source Anthyia. Published
with Anthyia's authorisation

has also revalued ramie with new products like threads and fabrics, among others. This company integrates and promotes this unusual fibre—ramie that, in turns, contributes to a legitimate corporate sustainable value-creation cycle.

A dilemma in the territory of sustainability is that the management systems and design principles that we have used to organize institutions are out of alignment with the underlying laws of nature and human nature. In turn, individual capacities seem to be the core of sustainable development: we see such leadership as necessarily going beyond conventional notions, because it needs to be able to step outside and challenge current formulations of society and business, and because sufficiently robust change means questioning the ground we stand on" (Marshall et al. 2011, p. 6). For this reason, the personal values and beliefs of Ying, together with innovation, have become Anthyia's focus.

Repositioning the future growth path of the company relies on Ying and her partners' abilities, competences, and technology used. Ramie research work is critical to this process. Overall, firms that make investments in these solutions resort to brand-new approaches to future issues and, in turn, use innovation for growth. Anthyia break-even point is not here yet, though the company has a clear vision of its future and sustainable model. Based on this, the company as a road-map with priorities, new products and technologies, and resources to be allocated.

Figure 3 depicts how the company builds sustainable value.

Sustainable value relies on a corporate leader and skill-set. This leader inspires people, in-house and in the outside world to work towards sustainability. Ying Luo's organisation is based on her vision of sustainability, as well as her value set and engagement. She is an inspiration to the team and the industry as well.

This leader's soul is embedded in the company. The same happens with new technology—both with ramie and knitting ramie products. Both technology and innovation contribute to value creation.

Ying's holistic managerial style with focus on opportunities, challenges, and a profoundly studied business environment contributes to the company's position going forward. To help in this process, the company's capacities and resources join the external world capacities and mandates. Diverse standpoints and strategies are put together from the internal and external realms. Anthyia is the combination of Ying's personal values and the Eastern perspective of ramie in a continuous creativity and innovation process. This is essential for the company's long-term existence and, since it is a dynamic concept, it requires managerial and organisational leadership that, as we have already noted, are typical of both Ying and Anthyia. In the case of the company, innovation is the core value to manage its destiny.

Openness creates a special trust to build reliable relationships with every stakeholder. Anthyia clearly shows this in the respect for the Eastern culture of

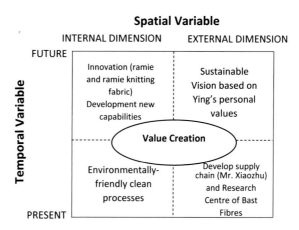

Fig. 3 Strategies and practices for creating sustainable value in Anthyis's future. *Source* Designed by the author (adapted from Stuart L. Hart and Mark Milstein, Creating Sustainable Value 2003)

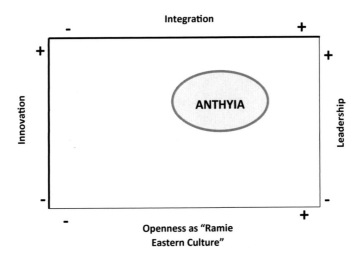

Fig. 4 Creating sustainable value at Anthyia: Relationship among Integration, Innovation, Leadership and Ramie Eastern Culture. *Source* Designed by the author

ramie, fostering a deep dialogue by being a good listener (when talking with the partners). In turn, Anthyia has a forward-looking view developing new products and new skills to create sustainable value.

As depicted in Fig. 4, convergence of integration, innovation and leadership, added to the extent of integration of the ramie Eastern culture, shows how Anthyia is creating sustainable value for the future and a competitive edge with a model focused on its partners and stakeholders' best interest.

References

Aiguo Z (unknown year) Ramie (*Boehmeria nivea*) Production and Its Diverse Uses in China. Presentation of the Institute of Bast Fibre Crops, Chinese Academy of Agricultural Sciences (IBFC, CAAS)

Berry CJ (1994) The idea of luxury—a conceptual and historical investigation. Cambridge University Press, New York

Carcano L (2013) Strategic management and sustainability in luxury companies: the IWC case. J Corp Citizensh 52:36–54

Cavagnaro E, Curiel G (2012) The three levels of sustainability. Greenleaf Publishing, Sheffield, UK

Cloutier D (2015) The vice of luxury—economic excess in a consumer age. Georgetown University Press, Washington

Choi YR, Gray ER (2008) The adventure development process of sustainable entrepreneurs. Manage Res News 31(8):558–569

Cohen AC, Johnson I (2014) Fabric science, 10th edn. Fairchild Books, New York

Dixon SEA, Clifford A (2007) Ecopreneurship: a new approach to managing the triple bottom line. J Organ Change Manage 20(3):326–345

Ehrenfeld JR (1999) Cultural structure and the challenge of sustainability. In: Sexton K, Marcus AA, Easter KW, Burckhardt TD (eds) Better environmental decisions—strategies for governments, businesses, and communities. Island Press, Washington

Featherstone M (2014) Luxury, consumer culture and sumptuary dynamics. J Luxury 1(1):47–70

Freeman RE (1984) Strategic management: a stakeholder approach. Pitman, Boston

Freeman RE, Martin K, Parmar B (2007) Stakeholder capitalism. J Bus Ethics 74:303–314

Gardetti MA (2011) Sustainable luxury in Latin America. Conference delivered at the seminar sustainable luxury & design within the framework of the MBA of IE- Instituto de Empresa-Business School, Madrid, Spain

Gardetti MA, Girón ME (2014) Sustainability luxury and social entrepreneurship: stories from the pioneers. Greenleaf Publishing, Sheffield

Gardetti MA, Girón ME (2016) Sustainability luxury and social entrepreneurship II: more stories from the pioneers. Greenleaf Publishing, Sheffield

Gardetti MA, Muthu SS (2015) Handbook of sustainable luxury textiles and fashion. Springer, London

Gardetti MA, Torres AL (2013) The Journal of Corporate Citizenship a Special Issue on Sustainable Luxury (Issue 52—December, 2013). Greenleaf Publishing, Sheffield

Gardetti MA, Torres AL (2014a) Sustainability Luxury: managing social and environmental performance in iconic brands. Greenleaf Publishing, Sheffield

Gardetti MA, Torres AL (2014b) Introduction. In: Gardetti MA, Torres AL (eds) Sustainability luxury: managing social and environmental performance in iconic brands. Greenleaf Publishing, Sheffield

Gholami S (2011) Value creation model through corporate social responsibility. Int J Bus Manage 6(9):148–154

Hamilton RW (2007) Bast and leaf fibers in the Asia-Pacific Region: an overview. In: Hamilton RW, Milgram BL (eds) Material choices—refashioning bast and leaf fibers in Asia and the Pacific. Fowler Museum at UCLA, Los Angeles

Hart SL (1995) A natural-resource-based view of the firm. Acad Manag Rev 20(4):986–1014

Hart SL (1997) Beyond greening—strategies for a sustainable world. Harvard Bus Rev 75(1): 66–76

Hart SL (2005) Capitalism at the crossroads. Wharton School Publishing, Upper Slade River

Hart SL (2007) Capitalism at the crossroads—aligning business, earth, and humanity, 2nd edn. Wharton School Publishing, Upper Slade River

Hart SL, Milstein M (1999) Global sustainability and the creative destruction of industries. MIT Sloan Manage Rev 41(1):23–33

Hart SL, Milstein M (2003) Creating sustainable value. Acad Manag Exec 17(2):56–67

Heine K (2011) The concept of luxury brands. Technische Universität Berlin, Department of Marketing, Berlin

Hoffman J, Lecamp L (2015) Independent luxury—the four innovation strategies to endure in the consolidation jungle. Palgrave Macmillan, New York

Kadolph SJ (2011) Textiles, 11th edn. Prentice Hall, Upper Saddle River

Kapferer JN (2015) Kapferer on luxury. Kogan Page, London

Kapferer JN, Basten V (2010) The luxury strategy—break the rules of marketing to build luxury brands. Kogan Page, London

Kleanthous A (2011) Simple the best is no longer simple. In: Raconteur on sustainable luxury 3 July 2011, http://theraconteur.co.uk/category/sustainability/sustainable-luxury/. Access on 7 Dec 2012

Kozlowski R, Rawluk M, Barriga-Bedoya J (2005a) Ramie. In: Franck RR (ed) Bast and other plant fibres. Woodhead Publishing and The Textile Institute, Cambridge

Kozlowski R, Rawluk M, Barriga-Bedoya J (2005b) Bast fibres (flax, hemp, jute, ramie, kenaf, abaca). In: Blackburn RS (ed) Biodegradable and sustainable fibres. Woodhead Publishing and The Textile Institute, Cambridge

Lips-Wiersma M, Morris L (2011) The map of meaning—a guide to sustaining our humanity in the world of work. Greenleaf Publishing, Shefield

Low T (unknown year) Sustainable luxury: a case of strange bedfellows. University of Bedfordshire, Institute for Tourism Research, Bedfordshire

Maisonrouge KP-S (2013) The luxury alchemist. Assouline Publishing, New York

Marshall J, Coleman G, Reason P (2011) Leadership for sustainability: an action research approach. Greenleaf Publishing, Sheffield, UK

Porter M, Kramer MR (2006) Strategy and society: the link between competitive advantage and corporate social responsability. Harvard Bus Rev: 1–15 (reprint)

Porter M, Kramer MR (2011) Creating share value. Harvard Bus Rev 1–17 (reprint)

Quinn R (1996) Deep change: discovering the leader within. Jossey-Bass, San Francisco

Rainey DL (2004) Sustainable development and enterprise management: creating value through business integration, innovation and leadership. article presented at Oxford University at its colloquium on 'Regulating Sustainable Development: Adapting to Globalization in the 21st Century', August 8 through 13, 2004

Shamina Y (2011) A review of the main concepts of luxury consumer behavior and contemporary meaning of luxury. Paper presented on the 2ème colloque franco-tchèque «Trends in International Business» Prague, 30 June 2011

Suzuki D, Dressel H (2002) Good news for a change: how everyday people are helping the planet. Stoddart Publishing Co., Toronto

Wittign MC, Sommerrock F, Beil P, Albers M (2014) Rethinking luxury—how the market exclusive products and services in an ever-changing environment. Roland Berger, London

World Commission on Environment and Development—WCED (1987) Our common future. Oxford University Press, Oxford

Young W, Tilley F (2006) Can business move beyond efficiency? The shift toward Effectiveness and equity in the corporate sustainability debate. Bus Strategy Environ 15(6):402–415

The Communication of Sustainability by Italian Fashion Luxury Brands: A Framework to Qualitatively Evaluate Innovation and Integration

Fabrizio Mosca, Chiara Civera and Cecilia Casalegno

Abstract Under the contemporary market conditions, Corporate Social Responsibility is required to turn into an integrated strategy for innovating and creating new business models. Luxury players are starting to embrace these strategies leveraging on many of their intrinsic characterizations that are, in fact, related to the concepts of sustainability and responsibility. Along with this tendency, also the communication of CSR is evolving in its contents and purposes and luxury brands are setting up strategies to engage stakeholders and customers on a plurality of communication means. Implementing an evolutionary approach of CSR requires greater changes inside the company and the communication of it needs to be homogeneous, transparent and, therefore, we argue, strategic. The power of luxury brands in the marketing expenditures favours their communication approach and impact, but what does it mean for a luxury player communicating about CSR and sustainability in a strategic way? Which are the boundaries between marketing and effective sustainable-driven strategy and communication? The chapter has twofold aims and uses a mixed methodological approach. Firstly, it aims at proposing a framework for qualitatively measuring the communication of CSR in an innovative strategic way. Secondly, it seeks to preliminarily investigate the extent of strategic CSR communication among a sample of 30 Italian fashion luxury players, through the content analysis of the CSR communications spread online, which is nowadays a trend that luxury market needs to constantly face. The research provides preliminary insights to the Italian fashion luxury brands about how to set up their strategic CSR communication and design a research model that can therefore be applied to a broader number of luxury brands.

F. Mosca (✉) · C. Civera · C. Casalegno
Department of Management, University of Turin, Turin, Italy
e-mail: fabrizio.mosca@unito.it

C. Civera
e-mail: chiara.civera@unito.it

C. Casalegno
e-mail: cecilia.casalegno@unito.it

© Springer Nature Singapore Pte Ltd. 2018
M. A. Gardetti and S. S. Muthu (eds.), *Sustainable Luxury, Entrepreneurship, and Innovation*, Environmental Footprints and Eco-design of Products and Processes, https://doi.org/10.1007/978-981-10-6716-7_5

Keywords Corporate social responsibility · Sustainability · Strategic CSR communication · Online CSR communication · Fashion and luxury

1 Introduction

Nowadays, Corporate Social Responsibility (CSR) has attracted much research attention (Vallaster et al. 2012; Visser 2012; Freeman et al. 2010), because both companies and stakeholders are focusing on sustainability in everyday business activities. The actual meanings and implications of CSR belong to and vary depending on industrial sectors, markets, and geographical contexts, and they are to be understood from more traditional to more innovative and integrated logics (Freeman et al. 2010). CSR practices sit at the core of corporate strategic priorities (Sen and Bhattacharya 2001), and so does the communication function, because every organization constantly talks to its stakeholders, and simultaneously, stakeholders constantly talk to each other about the firm's actions (Pastore and Vernuccio 2008). Communication, therefore, represents a strategic weapon to protect the firm against every kind of potential external and internal attack. Because the concepts of CSR and communication affect the whole organization—such that they are now recognized as a firm's strategic functions—as well as its reputation, it is very interesting to consider the extent to which they interact with each other in different markets, considering that the boundary separating commercial and other kinds of communication are disappearing (Casalegno and Civera 2016). At the same time, it is important to note that many firms use CSR messages to enhance their marketing communication to better market products and services (Castaldo et al. 2009) and increase competitiveness (Hur et al. 2014). Of course, it is also possible to analyze situations in which CSR interacts with the marketing function to lend a more ethical design to goods, services, processes, and policies to be shared with the target audience. As stated by most authors, the actions and policies implemented by firms always have a certain impact on the territory in which they are and on the society with which they interact (Torres et al. 2012; Freeman et al. 2010; Visser 2010, 2012; Carrol 2008; Freeman et al. 2006; Porter and Kramer 2002, 2011). Finally, CSR and the communication function need to interact with each other in a way through which it is also easy to satisfy marketing, reputational, and social requirements (Hur et al. 2014; Pomering and Dolnicar 2009).

Over the last three decades, companies' missions, activities, and business terminology and language clearly shift toward the inclusion of more sustainable, ethical, responsible, and environment-friendly practices, concerns, and values. It would be interesting to determine the extent to which this new attitude truly represents a shift toward a new way of doing business that might imply the renovation and redesign of business models and long-term strategies of the company. That is why, in a contemporary debate over CSR, it is necessary to consider its concrete meaning for companies, societies, and stakeholders and to understand its strong connection with all the strategic functions operating in a business organization.

Accordingly, CSR is a matter of strategy implementation, innovation, products, service and process redesign, marketing, communication, operations, human resources management, finance and investor relations, supply chain, lifecycle assessment, stakeholders' perception and consumers' preferences, community, environment, and society. In other words, to achieve its core purposes, CSR needs to be embedded in the organization as a whole, and it must afford economic, social, and environmental benefits at the same time.

The present research seeks to identify the contemporary characterizations of CSR in relation to communication for Italian luxury fashion brands and to measure the extent to which they implement strategic CSR communication.

The remainder of this paper is organized as follows. The first section presents the premise of this study, justifies the market and country chosen, and explains why the analysis of CSR communication is focused on online communication only. The second and third sections review literature on contemporary CSR and the relation between CSR and communication, and they end with the authors' elaboration of a framework for strategic CSR communication. The Methodology section explains the methodological approach used in this study. This study analyzes a sample of 30 luxury fashion brands through the framework of strategic CSR communication to accomplish the research objective.

Based on the findings, the investigated brands are ranked in terms of how strategic their CSR communication strategy is. The managerial implications for Italian luxury fashion brands include adopting the best practices emerging from case analyses and turning their CSR programs and communication into strategic and integrated functions that will allow a stronger firm-consumer relationship based on social improvement and improved competitiveness.

2 Premises: Why Focus on the Italian Luxury Fashion Market and on Online CSR Communication

In recent years, information technology has influenced both consumer and luxury goods markets. This influence has changed the way in which brands with high symbolic value communicate and distribute their products and share their brand value with communities. Luxury firms have found how to enhance their communication about their value proposition (Rifkin 2000) and overcome the so-called digital challenge (Mosca et al. 2013; Aiello and Donvito 2005), and consequently, they are changing how they communicate their product offering. Luxury firms are not just using digital and social networks to create a deeper link with their audiences; they have also started sharing news about their values and activities in the spheres of ethics, sustainability, environment, and responsibility (Mosca et al. 2013). This is a very interesting phenomenon and scholars are increasingly researching on the characterizations and the consequences on stakeholder perception of CSR communication in luxury markets. Moreover, along with the changing tendencies of companies to adopt online channels to communicate and sell, new consumptions mechanisms have emerged (Rifkin 2000). Consumers of luxury markets in geographically mature markets are more willing to participate in the process of sharing companies' values through online platforms (Mosca et al. 2013). This is why studies have analyzed the online communication of CSR by luxury fashion brands.

A literature review of luxury markets and analyses of some of the latest strategies of the largest luxury brands show that the concepts of responsibility and sustainability have gained increasing attention even in luxury markets during the last few decades, in spite of paradoxes and criticisms (Janssen et al. 2014; Kapferer 2010; Bendell and Kleanthous 2007; Vigneron and Johnson 1999). On the other hand, as stated previously, one can also identify the amount of attention luxury brands are devoting to the Internet and its use (Mosca et al. 2013, 2016; Kim and Ko 2012; Okonkwo 2009, 2010) owing to the need for value and brand content democratization. Luxury players have started considering online channels to communicate their efforts and projects concerning their responsibility and sustainability. This could increase public awareness of their intrinsic sustainable values, beliefs, and ethical concerns (Mosca et al. 2016; Eberle et al. 2013; Lee et al. 2013; Kaplan and Haenlein 2010; Insch 2008; Lodhia 2006).

Previous researches (Mosca et al. 2016) have shown that implementing CSR strategies and communicating them online is becoming an imperative for luxury brands, and this is an important trend to follow for two reasons. The first reason concerns the assumption that luxury and CSR present some similar principles (Janssen et al. 2014), because both are intended to convey strong values, craftsmanship, and a close link to the territory in which the brand is born. By definition, luxury goods and services are commonly considered rare and scarce, and this conveys a message of responsible consumption and protection of resources that luxury bases its character upon; the high dependence on resources results in a sort of obsession for sustainability (Kapferer 2010). Moreover, in the fashion industry, we consider sustainability and responsibility in the use of materials, implementation of processes, and management of value chain; these appear to be objects of criticism for wider audiences, and therefore, they are crucial in terms of the CSR perception. The second reason involves the abovementioned need of luxury brands for democratization, because they want to address a deep change affecting customers: on the one hand, polarization, and on the other hand, the changed reason for why they buy products with high symbolic value. Indeed, nowadays, customers want to participate in the creation of content that can be spread online; they want to feel closer to companies' values, and they love to recognize themselves in sustainable and responsible behaviors (Boston Consulting Group 2015; Lundquist 2014; Mosca et al. 2013; Okonkwo 2009; Rifkin 2000). In this light, it is easy to understand that for luxury players, the attempt to find a balance between the elusive intrinsic features of luxury goods/services and the necessity for a larger online consumer base is complex, and it can be partially solved by the creation of a strategic CSR communication plan.

This study focuses on Italian luxury fashion brands; these can be considered global players with local production. At first glance, this particularly affects their commitment to local communities.

3   CSR as Catalyst for Redesigning Business Models

Since the early 1990s, CSR has strengthened its managerial implications (Carrol 2008) and its relevance for companies' actions to include the typical concepts and concerns of sustainability, ethics, and environment into business operations and communication we well as concrete practices to implement, develop, and spread inside and outside the business organization (Vallaster et al. 2012; Freeman et al. 2004; Freeman and Phillips 2002; Macleod 2001; Mohr et al. 2001). Therefore, CSR has turned into an activity with increasingly concrete repercussions for both society and the company itself (Porter and Kramer 2011), and measuring CSR qualitative and quantitative results and performances has become increasingly required (Harrison and Freeman 1999).

Among all characterizations that CSR has been progressively acquiring over the last two decades, the focus on stakeholders' needs and expectations to reach their engagement represents the main driver for CSR policy, which has to be strategic in its purposes, contents and, communication (Lopez-De-Pedro and Gilabert 2012; Torres et al. 2012; Neal and Cochran 2008; Waddock and Bodwell 2007; Kotler and Lee 2005; Freeman et al. 2004)

Under such logic, the integrated CSR approach proposed by Freeman et al. (2010) moves beyond the traditional mentality of managers who have always considered a trade-off between profitability and responsibility (Waddock and Smith 2000) and adopt an innovative logic that allows players to combine social and environmental benefits with economic returns for the company and society at the same time. Some of the key contemporary literature reported in Table 1 below underline how CSR is considered a catalyst for the development of new sustainable business models, products, and processes.

According to the reported theories, implementing an integrated CSR strategy means that companies should conduct all CSR activities strategically by interacting with business functions across the organization to serve the interests of society and support business growth at the same time: For example, corporate philanthropy needs to be strategic (Porter and Kramer 2002) in the connection between charitable contributions and the company's core strategy, and it should have a positive impact on the corporate brand and image and, ultimately, equity, for instance, through cause-related marketing and/or strategic partnerships and donations (Maple et al. 2015; Civera and Maple 2012; La Cour and Kromann 2011; Liu and Ko 2011; Maple 2008; BITC 2004). Moreover, business repositioning and creation of new products and services that are sustainable, ethical, and responsible in their value chain, processes, components, materials, and objectives are the most effective

Table 1 Contemporary literature on CSR as catalyst for creating new business models

Authors	Conceptual framework
Freeman et al. (2010)	Integrated CSR is a part of core management concepts and processes, and it represents an integration of economic, social, ethical, and environmental decision-making criteria for corporate strategy.
Pauli (2010)	CSR integrates the reference community, environment, and companies' own resources to generate new businesses from waste and, in turn, opportunities for growth along the whole value chain.
Visser (2012)	- Systemic CSR is intended to innovate companies' business models and strategies by rethinking and redesigning processes, products, and services to optimize outcomes for a larger human and ecological system. - CSR 2.0 represents an integrated vision of responsibility and sustainability in a multi-stakeholder perspective that transforms the company and its partners (throughout the whole value chain) in a responsible system by developing innovative and sustainable processes, products, and services.
Crane et al. (2013)	CSR imposes social and economic alignment with a multi-stakeholder orientation, and it goes beyond the implementation of a set of values to develop concrete practices.

Source Authors' literature review

responses to real consumers' and stakeholders' needs (Candelo et al. 2015; Visser 2012; Small 2011; Pauli 2010; Macleod 2001; Mohr et al. 2001).

The positive repercussions of implementing this type of integrated CSR can be seen at different levels: improved work environment; increased employee productivity; reduced costs connected to waste management processes, lower carbon emissions, and energy saving; resource rationalization throughout the supply chain; reduced social and environmental costs; access to new markets and niche consumers through the development of innovative, sustainable, and ethical products and services; improved relationships with stakeholders in general; increased loyalty, trust, and positive brand perception from consumers; and enhanced corporate reputation (Candelo et al. 2015; Hur et al. 2014; Haanaes et al. 2011; Brammer and Millington 2005; Werther and Chandler 2005; Dawkins and Lewis 2003; McWilliams and Siegel 2001).

4 Online Communication by Luxury Brands

Communication has now become an integral part of luxury companies' brand strategies and, along with distribution, constitutes one of their two main drivers (Mosca 2010). Both the significant diffusion of the Internet and digital systems and a "more relaxed" relationship between customers and new technologies have deeply transformed the communication methods used by luxury companies over the last five years, with a progressive integration between traditional and new digital media (Trusov et al. 2009).

The sales point keeps its central position as the hub of communication activities; however, it integrates new technologies, thus allowing for pre- and post-sales customer service and an enhancement of the whole consumer experience (Linda 2010). User-generated content (UGC) is the most significant recent innovation. UGC guarantees a closer and more direct company-customer relationship and provides consumers with a completely new experience in relation to a brand with high symbolic value. The continuous increase in social networks is due to characteristics of these platforms, like easy accessibility and management of one's own social networks and their ability to attract millions of users (Gensler et al. 2013).

In general terms, these features identify social networks as mass tools, making them the exact opposite of anything that has to do with luxury and its offer (exclusivity, uniqueness, etc.). There are several prejudices in the use of social networks (Okonkwo 2010). Of these, two can be particularly dangerous for luxury brands:

– Social networks are mass tools. Almost all social networks are open to everyone. However, some online networks base their higher status on some sort of exclusivity (Tuten and Solomon 2014). Some social networks like Facebook (Dhaoui 2014) are defined by their huge number of users, whereas others like A Small World stand out for the quality of their community members. For example, 90% of A Small World users have an average income of 330,000$ (the remaining 10% is made up of millionaire philanthropists and aristocratic heirs). A Small World is therefore one of the online communities gathering the highest number of rich people on one platform in the world.
– Higher-income people do not use social networks. This statement is not accurate, as shown by several studies.[1] Rich people are very active on LinkedIn, A Small World, Twitter, and MySpace (for example, pop stars and their followers), and they love blogs through which they can chat with their virtual friends about sailing, flying, photography, and fashion (Toubia and Stephen 2013). Social networks based on community and trust provide people with a sense of belonging and affiliation.

Communication in luxury markets is evolving according to three main trends that are responsible for a change in consumer behavior:

– Search for personal experience and gratification
– Pervasiveness of technology and development of digital channels
– Corporate Social Responsibility (CSR)

Search for personal experience and gratification. Several studies have highlighted how experience is nowadays a key element for consumers, who look for it even before owning the purchased good. This is due to a change in consumers'

---

[1] A recent study carried out by the Luxury Institute showed that 72% of wealthy Americans belong to at least one social network. Luxury Institute Wealth Survey, *Social Networking Habits and Practices of the Wealthy*, 2009, on www.luxuryboard.com.

purchasing patterns. Consumers are now shrewd, well-informed, experience-hungry, in search of a connection with the brand, powerful, influential, and individualistic; however, at the same time, they are eager to be part of a community they can identify with. Therefore, communication activities become strategic and are integrated with distribution activities in the competitive market of luxury goods. Distribution gives value to a product's and brand's heritage and guarantees a remarkable purchasing experience.

Pervasiveness of technology and development of digital channels. Innovations have caused a change in consumers' purchasing behavior and role, from passive customers to active users. Customers are now willing to create content and be part of good production processes. They are involved in consumption (See-To and Ho 2014) and in a process of value co-creation. The development of new technologies and compression of product lifecycles are at the center of new consumption mechanisms that involve sharing goods and experiences more than buying products to possess them (Rifkin 2000). Digital channel communications help luxury companies understand the role of the Internet. In fact, the Internet is a communication medium among others as well as a multichannel platform that should be used along with other tools in specific integrated marketing strategies. Communication activities through digital media aim at building consumers' trust, providing information, and sharing and exchanging opinions and content in a circular process by integrating traditional unidirectional means of communication typical of press, radio, and TV.

CSR. In luxury markets, consumers are interested in companies' codes of ethics and conduct as well as the implementation of sustainable practices throughout the integrated value chain. Consumers expect the highest-quality standards as well as ethical decisions in relation to raw materials, labor conditions, relationships with staff (Hajli 2014), and selection of suppliers. Companies' strategic management should therefore pay attention to all these elements, both in their implementation and in coherent and integrated communication (Mosca et al. 2016).

5   Measuring the Communication of Sustainability: CSR Communication Framework

Given that the engagement of companies in CSR activities needs to become more integrated and strategic, the communication of these achievements and engagements to their stakeholders also has to be set up accordingly (Lllia et al. 2013). Literature and global marketing researches suggest that CSR communication is crucial for obtaining positive feedback from all stakeholder groups in terms of perception and trust as well as from the target market, which comprises actual and potential consumers who will be more likely to approach the brand and its selling proposition (Vallaster et al. 2012; Matute-Valejo et al. 2011; Pomering and Dolnicar 2009; Sen and Bhattacharya 2001). It is consumers' experiences that define corporates' reputation (Uwins 2014), and their attitudes and purchasing intentions driven by CSR—if consumers are aware of them (Pomering and Dolnicar 2009)—will ultimately impact the economic and image returns for companies (Demetriou et al. 2009; Brammer and Millington 2005). Accordingly, companies are steadily increasing their CSR communication; this is evident from the fact that CSR reporting has become a "business practice worldwide" (KPMG 2013) and that it is undertaken by most of the largest companies globally (Frostenson et al. 2011).

In a context in which consumers form their opinions differently than in the past, as they have continuous access to companies' information and behaviors (Christodoulides et al. 2011) beyond those directly communicated and managed by the company itself to affect their preferences and perception, we argue that CSR communication must become strategic.

In its conceptual definition, strategic CSR communication is a transparent, homogeneous, coherent, and integrated way of reporting on ethics, sustainability, responsibility, and environment. It establishes a multi-stakeholder and circular dialogue with internal and external clients and all stakeholders, from whom it takes feedbacks and to whom it is entirely addressed. Concretely, strategic CSR communication reports on the alignment between CSR promises and concrete actions undertaken; it is focused on informing rather than persuading, and that is why the communication of performances and concrete results and achievements is preferred to merely statements of values and promises (Hur et al. 2014; de Ven 2008).

Clearly, for CSR communication to be set up strategically, CSR actions and activities have to be conducted as integrated strategies, as shown in Fig. 1. The table shows the authors' suggestion for a framework containing three different dimensions of CSR that the communication should report on in a transparent and coherent way. The framework of strategic CSR communication will serve as tool to evaluate the spread of strategic CSR communication.

The communication of Standards is intended to report on the alignment and/or implementation of national or international standards in the areas of environment, quality, ethics, human resources, and reporting systems. This is the most formal type of CSR communication, in which the company shows itself to be aligned with the minimum requirements (both compulsory laws and voluntary standards compliance) demanded by a certain industry or sector and adopted by most players. It is not a differentiating CSR communication strategy.

The communication of Strategic Philanthropy is aimed at informing the community, as the main target, and the whole stakeholder group about the company's involvement with its reference societies among which it operates by buying, selling, and/or producing. This communication highlights the company's efforts to support causes of social interest that are pursued by Third Sector Organizations or any other cause in the field of art, culture, and sport that are coherent with and relevant to its own core business. Strategically, the company will report on concrete results reached by its sponsorships, cause-related marketing activities, donations, partnerships, community projects, initiatives of the foundation, and employee volunteering.

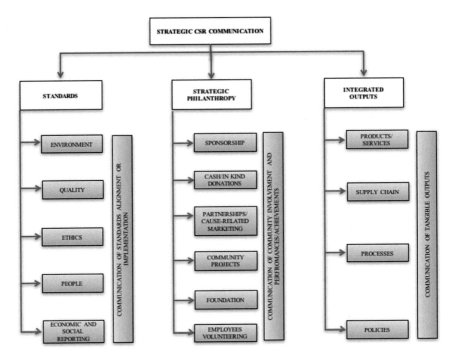

Fig. 1 Framework of strategic CSR communication. *Source* Authors' personal work

The communication of Integrated Outputs is set up to inform the audience of concrete achievements in the field of ethics, sustainability, environment, and responsibility. It is strictly linked to product and service development, supply chain management, and process and policy implementation. This is the dimension in which CSR embeds the organization at the core, and the communication of tangible performances is more effective because the outputs speak for themselves.

Of course, the more the actions and activities implemented within the three dimensions are integrated with the core business of the company, are coherent with each other, and express concrete achievements and tangible repercussions, the more their effective communication is facilitated, ultimately making it strategic. As explained above, for instance, the communication of sustainable products and services that have already been developed with sustainable and responsible criteria and/or are themselves an expression of sustainability (for example, Gucci's ecological shoes) increases the effectiveness of communication. In this specific case, CSR communication should be a natural output of the initiatives, with the aim of informing about tangible returns and concrete performances rather than merely convincing the audience about the intangible attitudes or values of the company.

To set up a more effective strategic CSR communication that is better perceived by consumers, companies need to be aware of which CSR communications influence their target market perception more effectively within a particular industry or market condition. This will help them leverage their communications on the implementation of CSR activities to increase loyalty and trust and, crucially, turn consumers' perceptions into positive brand associations.

6 Research Methodology

To provide some first insights into trends in luxury fashion brands' online strategic CSR communication, the case analysis methodology was used (Yin 2013). The analysis was based on the framework of strategic CSR communication explained previously. The aim of the analysis is to provide a ranking for measuring the adoption of online strategic CSR communication by the sample of Italian luxury fashion brands. The investigation is based on information and data obtained from the analysis of 30 cases by selecting the biggest Italian companies among those included in the Deloitte ranking "Global Powers of Luxury Goods Top 100" 2015, as shown in Table 2. This study examined both the online presence and the CSR contents that the brands published to share with their online audience. Information was collected over a 6-month period starting from September 2015 from the official websites of the luxury and fashion brands and from any other easily accessible online sources relevant to the subject at issue by using the principal online research engine and online buzz.

This research aims to further the discussion on the extent to which the brands of the considered sample communicate about their involvement in sustainability and responsibility issues. As a first step, the presence of an official CSR webpage and the chance to download the sustainability report from the web site was analyzed ("0" = they don't have, "1" = they have). Afterwards, the sample was tested through KPIs (or factors that are a part of each dimension of the framework) by adopting the framework of strategic CSR communication, as shown in Table 3. Moreover, each KPI was assigned a weight to obtain a final ranking (expressed as a mean) that shows which players communicate their CSR efforts most effectively and strategically. The total score for all three dimensions of strategic CSR communication (communication of standards, communication of strategic philanthropy, and communication of integrated outputs) can be between 0 and 1.

7 Findings and Discussions

7.1 Spread of Strategic CSR Communication

The analysis allowed, as a primary result, an estimation of the extent to which the brands adopt strategic online CSR communication, based on which KPIs they adopted, as shown in Table 4. Furthermore, Table 5 shows the details of the

Table 2 Sample of Italian luxury fashion brands

LVMH	FENDI
	BULGARI
	LOROPIANA
	EMILIO PUCCI
KERING SA	GUCCI
	BOTTEGA VENETA
	BRIONI
	SERGIO ROSSI
PRADA	
MIUMIU	
GIORGIO ARMANI	
OTB	DIESEL
	MARNI
MAX MARA	
ERMENEGILDO ZEGNA	
SALVATORE FERRAGAMO	
TODS'	TODS'
	HOGAN
	FAY
DOLCE & GABBANA	
MONCLER	
VALENTINO	
VERSACE	
BRUNELLO CUCINELLI	
ETRO	
LUIJO	
AEFFE	MOSCHINO
	ALBERTA FERRETTI
FURLA	
REPLAY	
ROBERTO CAVALLI	

Source Authors' personal work

ranking considering that the results were obtained by measuring each dimension of strategic CSR communication from 0 to 1 depending on the extent of the implementation of KPIs (or factors) that are a part of the considered dimension; therefore, the overall maximum score could be 3 out of 3. In particular, each KPI (Table 3) was weighted depending on the number of KPIs within each dimension. As shown in Table 4, Brunello Cucinelli had the highest score of 2.625; this indicates that this company had the highest spread of online strategic CSR communication.

These results show that some luxury brands do not communicate about CSR at all through their official website; instead, they refer customers to the CSR section of

Table 3 KPIs from the framework of strategic CSR communication

Communication of standards	Code of ethic/conduct
	Environmental standards
	Quality standards
	Employees standards
	Accounting standards
Communication of strategic philanthropy	Donations (third sector)
	Donation for art and culture
	Partnership 3rd sector
	Art and culture partnerships
	Sponsorship
	Employees volunteering
	Community projects
	Fundation
Communication of integrated output	Product
	Processes
	Sustainable supply chain
	Policy

Source Authors' personal work

Table 4 Ranking of Italian luxury fashion brands by online strategic CSR communication

CSR web com. mean	Firm	CSR web com. mean	Firm
2.625	BRUNELLO CUCINELLI	0.25	FURLA
2.3	PRADA	0.25	ROBERTO CAVALLI
2.15	GUCCI	0.2	DOLCE & GABBANA
1.925	BULGARI	0.125	FEIMDI
1.925	GIORGIO ARMANI	0.125	EMILIO PUCCI
1.75	TODS'	0.125	MARNI
1.425	MONCLER	0	BRIONI
1.35	VALENTINO	0	SERGIO ROSSI
1.25	LOROPIANA	0	MIUMIU
1.15	DIESEL	0	SALVATORE FERRAGAMO
1.125	ERMENEGILDO ZEGNA	0	HOGAN
0.875	MAX MARA	0	FAY
0.825	VERSACE	0	ETRO
0.7	BOTTEGAVENETA	0	LUIJO
0.5	REPLAY	0	ALBERTA FERRETTI
0.375	MOSCHINO		

Source Authors' personal work

Table 5 Italian luxury fashion brands' online strategic CSR communication—KPIs

		CSR section on the web site	Social/sustainability report	Communication standards mean	Communication of strategic philantropy mean	Communcation of integrated output mean
LVMH (social report del gruppo)	FENDI	0	1	0	0.125	0
	BULGARI	1	1	0.8	0.375	0.75
	LOROPIANA	0	0	0	0.25	1
	EMILIO PUCCI	0	0	0	0.125	0
KERING SA	GUCCI	1	0	0.4	0.75	1
	BOTTEGA VENETA	0	0	0.2	0	0.5
	BRIONI	0	0	0	0	0
	SERGIO ROSSI	0	0	0	0	0
	PRADA	1	1	0.8	0.75	0.75
	MIUMIU	0	0	0	0	0
	GIORGIO ARMANI	1	0	0.8	0.125	1
OTB	DIESEL	1	0	0.4	0.5	0.25
	MARNI	0	0	0	0.125	0
	MAX MARA	0	0	0	0.375	0.5
	ERMENEGILDO ZEGNA	1	0	0	0.625	0.5
	SALVATORE FERRAGAMO	0	0	0	0	0

(continued)

Table 5 (continued)

			Communication standards mean	Communication of strategic philantropy mean	Communcation of integrated output mean
TODS'	TODS'	1	1	0.5	0.25
	HOGAN	0	0	0	0
	FAY	0	0	0	0
	DOLCE & GABBANA	0	0.2	0	0
	MONCLER	0	0.8	0.375	0.25
	VALENTINO	0	0.6	0.25	0.5
	VERSACE	0	0.2	0.125	0.5
	BRUNELLO CUCINELLI	0	1	0.625	1
	ETRO	0	0	0	0
	LUUO	0	0	0	0
AEFFE	MOSCHINO	0	0	0.375	0
	ALBERTA FERRETTI	0	0	0	0
	FURLA	0	0	0.25	0
	REPLAY	0	0	0	0.5
	ROBERTO CAVALLI	0	0	0.25	0

Source Authors' personal work

the group they belong to (for instance, CSR information is communicated on the website of Tod's group but not on that of its product brand).

7.2 Clusters for Sustainability Communication

Table 6 shows the authors' clustering of luxury players' online CSR communication. In particular, cluster 1 is composed of web absent (they use the website to communicate about their products and brand heritage but not for CSR communication); cluster 2 by web challenged (they communicate mostly on Strategic Philanthropy without reporting the performances of their actions); cluster 3 by web friendly (they communicate mostly about Strategic Philanthropy by reporting on their promises and values, but they have increased communication about performances related to Integrated Outputs); cluster 4 by web influencers (they communicate mostly about performances related to Integrated Outputs). The results note that the use of CSR communication on websites was challenging for most brands in clusters 2 and 3.

From the above analysis of the investigated firms' CSR communication, the following results were found. Kering Group SA (Gucci, Bottega Veneta, Brioni, and Sergio Rossi) published information about its Code of Ethics online and stated that this code has been made mandatory since 2005. Moreover, it created a Group Ethics Committee that monitors how and why this Code is implemented to respond to various enquires from employees and to develop the Group's policies and actions

Table 6 Clusters of luxury players' online CSR communication

(1) Web absent	(2) Web challenged	(3) Web friendly	(4) Web influencers
ALBERTA FERRETTI	DOLCE & GABBANA	GIORGIO ARMANI	BRUNELLO CUCINELLI
BOTTEGA VENETA	EMILIO PUCCI	LOROPIANA	BULGARI
BRIONI	FENDI	MONCLER	DIESEL
ETRO	MARNI		ERMENEGILDO ZEGNA
FAY	MAX MARA		GUCCI
HOGAN	VERSACE		PRADA
LUIJO			VALENTINO
MIUMIU			
MOSCHINO			
SALVATORE FERRAGAMO			
SERGIO ROSSI			
TODS'			

Source Authors' personal work

in the area of sustainable development. Even if such communication is available online, the individual brands' websites do not contain information, and therefore, they are considered web absent. Bulgari, which is a web influencer owing to the heavy presence of online information on its sustainable and responsible actions, states its commitment to environmental issues by adopting sustainable practices within production and management activities. For instance, its environmental policy is based on the ISO14001 standard, and it is aimed at increasing the environmental awareness of its employees, partners, and suppliers. The aim is to meet concrete goals in waste management and in the use of materials and resources. Bulgari also considered the threat represented by energy consumption, and it claimed that it makes many efforts to increase sensitivity about sustainable forms of mobility. Another case to consider among web challenged brands is Fendi and the communication of its strategic philanthropy. Indeed, Fendi provided funds for the restoration of the Trevi Fountain in Rome in 2015, and its involvement in this project confirmed the Roman origins of this brand. Furthermore, the Carla Fendi Foundation is the brand's institutional foundation for conducting restoration projects such as that of the "Teatro Caio Melisso Spazio Carla Fendi" to achieve its mission of protecting cultural heritage and values from the past. Giorgio Armani is considered a web friendly brand; its management decided to implement a long-term sustainability policy for all stages of the supply chain to monitor and reduce the use of chemical components. The purpose of this policy, which was started in 2013, is to achieve a zero-discharge goal by 2020; it constantly conducts an audit program to monitor the progress of its efforts and their environmental impact.

8 Conclusions, Implications and Further Researches

This study was framed around the online communication of CSR, being both online communication and sustainability two of the most ongoing challenges and trends for luxury brands. Findings have shown that—majorly at a group level—Italian fashion luxury players have dealt with the process of luxury democratization through a strategic communication of CSR, intensifying the production of integrated outputs and the use of strategic philanthropy. Luxury goods are often perceived as wasteful and latest corporate scandals by top brands are pushing players to intensify the implementation of sustainable and responsible practices in their manufacturing more intensively and concretely and, to communicate to their stakeholders that greater attention is put on CSR and avoid the bad repercussions of the past. These results show that luxury brands have understood the importance of linking traditional marketing communication over sustainability with performances connected to actual results and concrete evidences of CSR implementation. Moreover, they shed light on the fact that integrated outputs are a means for sustainability and innovation to become key driver of luxury brands' core strategies. However, the findings suggest that luxury fashion players need to strengthen and/or implement their online communication strategically at a brand level. Specifically, they should report the sustainable performances of their products and processes as well as value chain management. The communication of integrated outputs and strategic philanthropy appears to be more effective as it shows both intentions and promises as well as tangible performances and repercussions on the environment and the community. These results open up interesting avenues for future researches on the merge between "online luxury" and "responsible luxury" in a more strategic way, so that products can become integrated outputs of communication about sustainability on integrated channels (social media, websites and points of sale for instance) and CSR can become more than just a policy; instead a confirmation and evidence of appealing unique goods and transparent luxury management strategies.

One of the main outputs of this study, which makes it highly replicable in any industry and geographical context, is the framework of strategic CSR communication that can serve as tool for qualitatively evaluating the extent to which CSR communication is strategically implemented by firms.

The topic of online strategic CSR communication represents an interesting area of research, that can be furthered by both extending the sample to luxury players operating in various industries and including an investigation of luxury customers' perception over online CSR communication. Moreover, the research could be broadened by including a sample of luxury companies from France and the United Kingdom to outline whether and which cultures and values affect luxury brands in the implementation and communication of CSR in a strategic and innovative way.

References

Aiello G, Donvito R (2005) Comunicazione integrata nell'abbigliamento: strategie di marca e ruolo del punto vendita nella distribuzione specializzata statunitense. Conference Le Tendenze del Marketing in Europa, Ecole Supérieure de Commerce de Paris ESCP-EAP, Paris 21–22 Jan 2005

Bendell J, Kleanthous A (2007) Deeper luxury: quality and style when the world matters. WWF-UK

Boston Consulting Group (2015) True luxury global consumer insight. Report at http://www.luxesf.com/wp-content/uploads/2014/06/bcg-altagamma-true-luxury.pdf

Brammer S, Millington AI (2005) Corporate reputation and philanthropy: an empirical analysis. J Bus Ethics 61(1):29–44

Business in the Community (BITC) (2004) Brand benefits: how cause related marketing impacts on brand equity, consumer behaviour and the bottom line. Report by BITC, October. www.bitc.org.uk

Candelo E, Casalegno C, Civera C (2015) Towards corporate shared value in retail sector: a comparative study over grocery and banking between Italy and the UK. J Econ Behav 5:105–120

Carrol AB (2008) A history of corporate social responsibility, concepts and practices. In: Crane A, Matten D, McWilliams A, Moon J, Siegel DS (eds), The Oxford handbook of corporate social responsibility. OUP Oxford

Casalegno C, Civera C (2016) Impresa e CSR: la "non comunicazione" di successo. Regole per una gestione responsabile delle relazioni. [CSR Communication: communicating better, communicating less] Increase Franco Angeli, Italy

Castaldo S, Perrini F, Misani N, Tencati A (2009) The missing link between corporate social responsibility and consumer trust: the case of fair trade products. J Bus Ethics 84(1):1–15

Christodoulides G, Jevons C, Blackshaw P (2011) The voice of the consumer speaks forcefully in brand identity: user-generated content forces smart marketers to listen. J Advert Res 51 (1):101–111

Civera C, Maple P (2012) The magic or myth of corporate philanthropy? Conference paper at ISM-Open Institute of Social Marketing Conference on "Taking Responsibility: Social Marketing and Socially Responsible Management". The Open University, UK

Crane A, Matten D, Spence LJ (2013) Corporate Social Responsibility: in a global context. In: Crane A, Matten D, Spence LJ (eds) Corporate social responsibility: readings and cases in a global context. Routledge, Abingdon

Dhaoui C (2014) An empirical study of luxury brand marketing effectiveness and its impact on consumer engagement on Facebook. J Glob Fashion Mark 5(3):209–222

Dawkins J, Lewis S (2003) CSR in stakeholder expectations: and their implication for company strategy. J Bus Ethics 44:185–193

Demetriou M, Papasolomou I, Vrontis D (2009) Cause-related marketing: building the corporate image while supporting worthwhile causes. J Brand Manage 17:266–278

De Ven B (2008) An ethical framework for the marketing of corporate social responsibility. J Bus Ethics 82(2):339–352

Eberle D, Berens G, Li T (2013) The impact of interactive corporate social responsibility communication on corporate reputation. J Bus Ethics 118(4):731–746

Freeman RE, Phillips RA (2002) Stakeholder theory: a libertarian defence. Bus Ethics Q 12 (3):333–349

Freeman RE, Wicks AC, Parmar B (2004) Stakeholder theory and the corporate objective revisited. Organ Sci 15(3):364–369

Freeman RE, Wicks A, Harrison J, Parmar B, de Colle S (2010) Stakeholder theory: the state of the art. Cambridge University Press

Freeman RE, Velamuri SR, Moriarty B (2006) Company stakeholder responsibility: a new approach to CSR, business roundtable. Institute for Corporate Ethics Bridge Paper

Frostenson M, Helin S, Sandstrom J (2011) Organizing corporate responsibility communication through filtration: a study of web communication patterns in Swedish retail. J Bus Ethics 100:31–43

Gensler S, Völckner F, Liu-Thompkins Y, Wiertz C (2013) Managing brands in the social media environment. J Interact Mark 27(4):242–256

Haanaes K, Balagopal B, Arthur D, Kong MT, Velken I, Kruschwitz N, Hopkins MS (2011) First look: the second annual sustainability & innovation survey. Mit Sloan Manage Rev 52(2)

Hajli MN (2014) A study of the impact of social media on consumers. Int J Mark Res 56(3):387–404

Harrison JS, Freeman RE (1999) Stakeholders, social responsibility, and performance: empirical evidence and theoretical perspectives. Acad Manage J Oct:479–485

Hur WM, Kim H, Woo J (2014) How CSR leads to corporate brand equity: mediating mechanisms of corporate brand credibility and reputation. J Bus Ethics 125(1):75–86

Insch A (2008) Online communication of corporate environmental citizenship: a study of New Zealand's electricity and gas retailers. J Mark Commun 14(2):139–153

Janssen C, Vanhamme J, Lindgreen A, Lefebvre C (2014) The catch-22 of responsible luxury: effects of luxury product characteristics on consumers' perception of fit with corporate social responsibility. J Bus Ethics 119(1):45–57

Kaplan AM, Haenlein M (2010) Users of the world, unite! The challenges and opportunities of social media. Bus Horiz 53(1):59–68

Kapferer JN (2010) Luxury after the crisis: Pro logo or no logo. Eur Bus Rev:42–46

Kim AJ, Ko E (2012) Do social media marketing activities enhance customer equity? An empirical study of luxury fashion brand. J Bus Res 65(10):1480–1486

Kotler P, Lee N (2005) Corporate social responsibility. Doing the most good for your company and your cause. Wiley, Hoboken

KPMG (2013) Survey of corporate responsibility, report

La Cour A, Kromann J (2011) Euphemisms and hypocrisy in corporate philanthropy. Bus Ethics: Eur Rev 20(3):267–279

Lee K, Oh WY, Kim N (2013) Social media for socially responsible firms: analysis of fortune 500's Twitter profiles and their CSR/CSIR ratings. J Bus Ethics 118(4):791–806

Lodhia SK (2006) Corporate perceptions of web-based environmental communication: an exploratory study into companies in the Australian minerals industry. J Account Organ Change 2(1):74–88

Lundquist (2014) Digital disruption and the future of CSR. Findings from the Lundquist CSR online awards survey 2014, report

Linda SLL (2010) Social commerce-e-commerce in social media context. World Acad Sci Eng Technol 72:39–44

Liu G, Ko WW (2011) An analysis of cause-related marketing implementation strategies through social alliance: partnership conditions and strategic objectives. J Bus Ethics 100:253–281

Lllia L, Zyglidopoulos SC, Romenti S, Rodriguez-Canovas B, de Valle Gonzales, Brena A (2013) Communicating corporate social responsibility to a cynical public. MIT Sloan Manage Rev 54 (3):2

Lopez-De-Pedro JM, Gilabert Rimbau E (2012) Stakeholder approach: what effects should we take into account in contemporary societies. J Bus Ethics 107:147–158

Macleod S (2001) Why worry about CSR? Strateg Commun Manage 5(5):8–9

Maple P (2008) The spectrum of philanthropy. Caritas London 5:34–36

Maple P, Civera C, Casalegno C (2015) An investigation of "the spectrum of corporate social responsibility." or to be more precise: over-communication—a comparative analysis of the UK and Italian banking sectors from the consumers' perspective. Proceeding from IS4IS Summit Vienna June 2015

Matute-Vallejo J, Bravo R, Pina JM (2011) The influence of corporate social responsibility and price fairness on customer behaviour: evidence from the financial sector. Corporate social responsibility and environmental management 18 n/a. doi:10.1002/csr.247

McWilliams A, Siegel AD (2001) Corporate social responsibility: a theory of the firm perspective. Acad Manag Rev 26(1):117–127

Mohr BA, Webb DJ, Harris KE (2001) Do consumers expect companies to be socially responsible? The impact of corporate social responsibility on buying behavior. J Consum Aff 35(1):45–72

Mosca F (2010) Il marketing e le innovazioni del digitale. In: Re p, Mosca F, Bertoldi B (Eds) 'Marketing e nuovi scenari competitivi. Strategie e creazione di valore nella relazione con il cliente' McGraw-Hill Education

Mosca F, Casalegno C, Feffin A (2013) Nuovi modelli di comunicazione nei settori dei beni di lusso: un'analisi comparata. Paper X Convegno della Scuola Italiana di Marketing, Milano, 7–9 Oct 2013

Mosca F, Civera C, Casalegno C (2016) Luxury and corporate social responsibility communication strategies. How much does the web matter? A cross investigation on players and consumers' perception. Books of proceedings BAM (British Academy of Management) Conference, 6–8 Sept 2016, Newcastle University, UK

Neal R, Cochran LP (2008) Corporate social responsibility, corporate governance, and financial performance: lesson from finance. Bus Horiz 51:535–540

Okonkwo U (2009) Sustaining the luxury brand on the internet. J Brand Manage 16(5–6):302–310

Okonkwo U (2010) Luxury online: styles, systems, strategies. Springer

Pastore A, Vernuccio M (2008) Impresa e comunicazione. Principi e strumenti per il management. Apogeo Editore

Pauli G (2010) The blue economy: 10 years, 100 innovations, 100 million jobs. Paradigm Publications

Pomering A, Dolnicar S (2009) Assessing the prerequisite of successful CSR implementation: are consumers aware of CSR initiatives? J Bus Ethics 85(2):285–301

Porter M, Kramer M (2002) The competitive advantage of corporate philanthropy. Harvard Bus Rev 80(12):57–68

Porter M, Kramer M (2011) Creating shared value. How to reinvent capitalism and unleash a wave of innovation and growth. Harvard Bus Rev 89(1/2):62–77

Rifkin J (2000) L'era dell'eccesso. Milano, Mondadori

See-To EW, Ho KK (2014) Value co-creation and purchase intention in social network sites: the role of electronic word-of-mouth and trust–a theoretical analysis. Comput Hum Behav 31:182–189

Sen S, Bhattacharya CB (2001) Does doing good always lead to doing better? Consumer reactions to corporate social responsibility. J Mark Res 38(2):225–243

Small L (2011) Corporate responsibility, marketing and economic development in emerging markets. www.accountability.org/about-us/news/accountability-1/accountability.html

Toubia O, Stephen AT (2013) Intrinsic vs. image-related utility in social media: why do people contribute content to twitter? Mark Sci 32(3):368–392

Torres A, BIjmolt T, Tribò J, Verhoef P (2012) Generating global brand equity through corporate social responsibility to key stakeholders. Int J Res Mark 29(1):13–24

Trusov M, Bucklin RE, Pauwels K (2009) Effects of word-of-mouth versus traditional marketing: findings from an internet social networking site. J Mark 73(5):90–102

Tuten T L, Solomon M R (2014) Social media marketing. Sage, Thousand Oaks

Uwins S (2014) Creating loyal brands. A guide to earning loyalty in a connected world, Publish Green

Vallaster C, Lindgreen A, Maon F (2012) Strategically leveraging corporate social responsibility: a corporate brand perspective. Calif Manag Rev 54(3):34–60

Vigneron F, Johnson LW (1999) A review and a conceptual framework of prestige-seeking consumer behavior. Acad Mark Sci Rev 1

Visser W (2010) CSR 2.0: The evolution and revolution of corporate social responsibility. Responsible business: how to manage a CSR strategy successfully, pp 311–328

Visser W (2012) The Future of CSR: Towards Transformative CSR, or CSR 2.0. Kaleidoscope Futures Paper Series 1

Waddock S, Smith N (2000) Corporate responsibility audits: doing well by doing good. Sloan Manag Rev 42(2):75–87

Waddock S, Bodwell C (2007) What is responsibility management? And why bother? In: Total responsibility management: the manual, hardback, 9–25

Werther B, Chandler B (2005) Strategic corporate social responsibility as global brand insurance. Bus Horiz 48:317–324

Yin RK (2013) Case study research: design and methods. Thousand Oaks, Sage publications

The Relevance of Sustainability in Luxury from the Millennials' Point of View

Marius Schemken and Benjamin Berghaus

Abstract Over the past decades, aggravating environmental and social conditions on planet earth, partially caused by economic prosperity and correlating factors, have urged people to think about sustainability in all its facets and act accordingly. Companies, as a vital part of the economy, are pressured by legislative requirements and changing consumer needs. This shift in consumer behaviour is mainly embodied by the millennial generation who purchase more consciously and care more about ethical and environmental aspects than previous generations. Scholars previously studied the impact of Corporate Social Responsibility (CSR) on luxury firms and their image. However, the analysis of the millennials' point of view in this regard is still an uncared field of research. This gap is addressed by the following study. Luxury products are predestined to pioneer regarding sustainability based on their durability, quality excellence, and emotional value. Thus, the luxury goods industry cannot escape this development and thus must react. This paper investigates, by means of a focus group interview with millennials, in which respect the image of luxury firms from the millennial generation's point of view is influenced by sustainability efforts. The paper concludes that millennials have manifold and ambivalent associations with sustainability and luxury, that apportioned synergies between sustainability and luxury are hardly captured by millennials, and that millennials discuss the matter differently based on their perspective.

Keywords Luxury · Sustainability · Millennial consumer
Millennial job seeker · Focus groups

List of Abbreviations

CSR   Corporate Social Responsibility
ibid   ibidem
LVMH Louis Vuitton Moët Hennessy

M. Schemken (✉) · B. Berghaus
Institute of Marketing, University of St. Gallen, St. Gallen, Switzerland
e-mail: marius.schemken@t-online.de

B. Berghaus
e-mail: benjamin.berghaus@unisg.ch

© Springer Nature Singapore Pte Ltd. 2018
M. A. Gardetti and S. S. Muthu (eds.), *Sustainable Luxury, Entrepreneurship, and Innovation*, Environmental Footprints and Eco-design of Products and Processes, https://doi.org/10.1007/978-981-10-6716-7_6

1 Introduction

The importance of sustainability has increased over the past decades (Pufé 2014). The widening gap between resource consumption by mankind and resource provision by nature presents itself as one of the major issues in the environmental discussion (Petersen 2012) and thus is one of many leverage points for companies to act sustainably (Ahrend 2016). The shift towards a more sustainable economy is amplified by changing consumer behaviour on the demand side (Gurtner and Soyez 2016). This is mainly driven by the influential millennial generation, which consists of people born between 1980 and 2000, who consume more consciously, ascribe more importance to sustainability than previous generations, and critically question existing paradigms (Barton et al. 2012; Eastman and Liu 2012; Hill and Lee 2012; Hwang and Griffiths 2017; Moroz and Polkowski 2016; Pomarici and Vecchio 2014). Their influence also challenges industries and how firms position themselves as employers to attract motivated and dynamic personnel (Leveson and Joiner 2014; Stewart et al. 2017). This mindset challenges luxury goods manufacturers and their image also regarding sustainability efforts (Barton et al. 2012; Mahler 2017). Many luxury goods firms already act sustainably on multiple levels of the value chain (Amatulli et al. 2017; Kering 2017b; LVMH 2017; Richemont 2016; Winston 2016; Wittig et al. 2014).

However, there are only few studies built upon insight derived from primary data that evaluate millennials' opinion about sustainability in luxury. It is claimed that millennials have a different awareness of environmental issues and demand corporations to act sustainably. Yet, an explorative view on millennials' true perception of sustainability's importance within luxury has not yet been attempted. Thus, luxury goods manufacturers can hardly capture what their future customers really desire. Consequently, the guiding questions of this study are:

(1) What are millennials' associations with the terms sustainability and luxury?
(2) How important is sustainability in luxury from the millennial point of view?
(3) What are the differences between millennials' consumer and job seeker perspective?

To find answers to these questions, this study follows an explorative approach by building upon focus group interviews. Two focus group sessions are conducted that give insights into how millennials assess intersections of sustainability and luxury from both the consumer and the job seeker point of view. This shall give luxury goods manufacturers insights into their standing as producers and employers.

Based on this approach, the key findings are that (1) millennials define the terms sustainability and luxury not only with great variety but also with contradictions, (2) the relevance of sustainability in luxury in the eye of millennials seems to be overstated, and (3) there is a difference between the argumentation from consumer and job seeker perspective.

In the following chapters, the three main theoretical topic areas are presented, interlinkages are established, and a synthesis for investigation is identified. Then, the research design is presented. The corresponding findings are brought into context, and concrete recommendations for luxury goods manufacturers to react to millennials' opinions on the importance of sustainability in luxury are generated.

2 Theory

Within the scope of this study, three main research areas are of vital significance. First, the general concept of sustainability and its increasing importance for the business industry challenging current business models and practices in the light of drastically changing environmental conditions. Second, the contemporary understanding of luxury is influenced by these circumstances freeing up space for interpretation. Third, the millennial generation as a phenomenal example of a generation with a significantly different mindset than previous generations challenge existing paradigms. In the intersection of these three aspects lies the purpose of this study.

2.1 Sustainability

The contemporary understanding of the concept of sustainability mainly derives from two fundamental works (Buerke 2016): *Club of Rome*'s report "The Limits to Growth" (1972) states that—under the condition of a growing world population, an industrialised production, and an increasing consumption of natural resources—the natural boundaries of economic growth would be reached within a century. The authors, however, saw a chance to positively change these trends claiming that a long-term beneficial balance between ecological and economic factors would ensure the life basis for mankind permanently (Meadows et al. 1972). "Our Common Future" written by the *World Commission on Environment and Development (WCED)* establishes a general definition of sustainable development that "is development, which meets the needs of the present without compromising the ability of future generations to meet their own needs" (WCED 1987 p. 43). Since then, major catastrophes, such as the blowout of the oil platform *Deepwater Horizon* in 2010 ("Timeline: BP Oil Spill" 2010) and the core melt accident at the nuclear power plant in Fukushima, Japan in 2011 ("Japan earthquake: Explosion" 2011) drastically raise the awareness for a holistic approach to eliminate problems before they occur.

Therefore, an integrative approach is necessary to discuss sustainability in all its facets and to keep the system as a whole in an advantageous balance (Petersen 2012; Pufé 2014). This basically regards the three elements of the *Triple-Bottom-*

Line, which comprises the dimensions of economic, ecological, and social sustainability (Pufé 2014). According to this triad, organisations must realign their strategies and adapt their business models based on the premise of global justice by implementing common sense measures to transparently communicate to consumers whether a company acts sustainably (Ahrend 2016; Pufé 2014).

However, economic advantages of sustainable operations are not directly observable which weakens the power of incentives to behave sustainably. In the fashion and textile industry, for example, "there is a fear that adopting a sustainable mindset would deter innovation and creativity, seeing sustainable development as meaning frugality and minimalism" (Montesa and Rohrbeck 2014 p. 410). Further reasons for resistance to act sustainably are profit losses, complexity of execution, lack of personnel, insufficient political and societal support, security awareness, habit, and fear of change (Pufé 2014). Moreover, consumers oftentimes only perceive the Triple-Bottom-Line's ecological dimension as relevant (Martínez et al. 2014).

2.2 *Luxury*

There is no single or homogenous definition of what luxury is. Luxury is interpreted on an individual basis and thus obtains a variety of personal meanings (Kapferer and Laurent 2015). While some people may refer to luxury on a material level, others define their personal luxury by spending more time with family and friends. However, luxury products are often related to as superfluous because only affluent people can afford these goods (Windsor 2014). This also results from the fact that luxury products are costly and therefore scarce. Because this image also highly depends on the respective standpoint influenced by generation, zeitgeist, status, and societal environment (Amatulli et al. 2017; Wiedmann et al. 2007), thresholds in terms of price vary across generations, countries, and cultures (Cervellon and Shammas 2013; Kapferer and Laurent 2015). In this regard, Bastien and Kapferer (2013) distinguish between six meanings, of which three relate to the personal definition of luxury: product-related luxury, intangible luxury, and luxury as personal reward or social distinguisher. Luxury firms are known for producing goods based on exceptional craftsmanship, superior quality, and highly-trained personnel. Thus, respective products are made to last and to be passed on to following generations (Wittig et al. 2014). Luxury goods are also increasingly purchased due to impressive motives, such as quality, durability, and fit, instead of uniqueness or status factors (Cervellon and Shammas 2013; Hudders 2012). Consumers oftentimes perceive luxury brands through a distinctive product, for instance *Chanel* through their perfume *N° 5* and *Porsche* through the *911* (Kapferer 1998). The tangible value of such products, however, is significantly enhanced by an emotional value that is communicated by the luxury brand via extensive marketing (Seo and

Buchanan-Oliver 2015). Moreover, Amatulli et al. (2017) identify typical characteristics, which are attributed to luxury goods: quality excellence; scarcity, uniqueness, and exclusivity; aesthetic beauty; ancestral heritage and personal history; fulfilment of dreams; upper price range. These aspects contribute to consumers' brand and product perceptions.

However, current developments on the demand side raise the question whether the contemporary definition of luxury and the its epitome are challenged in fundamental aspects (Wittig et al. 2014).

2.3 Millennials

Most commonly, millennials are born between the years 1980 and 2000 ("Millennials coming of age" 2017; "Millennials outnumber baby boomers" 2015). They are also called digital natives and are attributed with high social consciousness as well as ecological awareness (Barton et al. 2012; Eastman and Liu 2012). The relevance of this generation's different mindset is substantiated by their presence and their spending power. In the United States of America alone, members of the millennial generation account for more than 25% of the population and accumulate USD 1.3 Trillion in direct spending (Barton et al. 2014). They spend their money primarily on products that embody their values and beliefs and thus represent an extension of their personality (Bergh 2017). A general shift in millennials' norms, attitudes, priorities, and expectations in comparison with previous generations can be observed (Kirchhof and Nickel 2014; Popovici and Muhcina 2015). Personal property, for instance, is increasingly perceived as a burden. Hence, sharing platforms for music, films, automobiles, flats, handbags, and even clothes are on the rise (Hwang and Griffiths 2017).

In many aspects, the characteristics of millennials represent a game changer and thus are a challenge for the respective industries, for instance in the fields of employee commitment towards a company (Stewart et al. 2017) and consumer behaviour (Gurtner and Soyez 2016). Due to their high engagement in social media channels, millennials influence not only each other but also companies and the society as a whole (Bolton et al. 2013). Regarding job preferences, literature states that millennials look for employer attributes, such as positive working environment, supportive culture, stimulating work that offers opportunities for advancement, long-term career progression, and variety in daily work routine (Guillot-Soulez and Soulez 2014; Hershatter and Epstein 2010). Moreover, a need for independence, equilibrated work-life balance, personal enjoyment, and the secondariness of salary characterise millennial job seekers (ibid.). Thus, expectations are rather related to individual than overarching issues (Ng et al. 2010).

2.4 Sustainability and Luxury

Although the luxury and apparel industry is one of the most environmentally harmful industries (Montesa and Rohrbeck 2014), and every so often a scandal is revealed ("Exposed: Crocodiles and alligators factory-farmed" 2017; "Slaving in the lap of luxury" 2013), the matter of sustainability does also concern these firms. Whilst some use lip service for reputation reasons, others develop and practice environmentally friendly ways to source and process their materials (Doval and Batra 2013). Despite the apparent contradiction between sustainability and luxury —opulence versus thrift, hedonism versus altruism, and superfluity versus necessity —luxury per definition in fact strives for sustainability (Amatulli et al. 2017; Winston 2016). Many luxury firms—owned by either *Louis Vuitton Moët Hennessy* (*LVMH*), *Kering,* or *Richemont*—pursue *Corporate Social Responsibility* (*CSR*) strategies and concretely communicate both their efforts and successes with sustainable products as well as their engagement in projects that increase attention towards the need for sustainability (Corporate Knights 2015, 2016; Cowdrey 2007; Kering 2017a, b; LVMH 2017; Richemont 2016; "Top green companies" 2016; Waller and Hingorani 2014; Wittig et al. 2014).

However, sustainability in the luxury industry predominantly still happens mostly unknown to consumers (Givhan 2015) because its open communication is not primarily pursued (Carcano 2013; Wittig et al. 2014). This may result from the fact that the green image of sustainable is not beneficial for luxury goods (Givhan 2015; "There is more to Kering's sustainability report" 2016).

2.5 Millennials and Luxury

In general, a distinct aspiration for luxury, in both intangible and product-related ways, can be attributed to the millennial generation, whereas each experience or product obtains a unique reward characteristic (Bulthuis 2015). In this context, Schade et al. (2016) establish that millennials' luxury brand purchase behaviour is impacted by (1) the social-adjustive function, which "is defined as a tendency to purchase and use brands to gain approval in social situations and to maintain relationships" (p. 316) by aligning with a peer group and (2) the value-expressive function, which "is defined as a tendency to purchase and use brands to communicate one's self-identity" (p. 316) and to represent personal characteristics (Bian and Forsythe 2012; Wiedmann et al. 2007). The individuality factor is corroborated by the fact that the more people in one's surrounding buy a respective luxury brand, the lower is the satisfaction of one consuming this brand (Shukla et al. 2016). Because millennials have been exposed to luxury goods in life earlier than their parents, they understand the sophistication luxury goods communicate and the exclusivity that comes along with that (Wittig et al. 2014). This also means that millennials have a distinct perception of certain brands that hardly changes over time (Kapferer 1998).

2.6 Sustainability and Millennials

Gurtner and Soyez (2016) state that millennials "want both to protect the environment and at the same time to enjoy the products they consume" (p. 105). They are more aware of environmental problems and consume more deliberately (Hill and Lee 2012; Moroz and Polkowski 2016; Pomarici and Vecchio 2014). Millennials care most about the sustainability of the products themselves, followed by sustainable manufacturing processes and social responsibility of the producing company (Mahler 2017).

However, this generation does not possess sufficient knowledge about the holistic concept of sustainability yet in order to use their awareness and take action accordingly (Hill and Lee 2012). Apart from that, millennials do know how to integrate sustainability into their daily routines by buying green and fair products, for instance.

On top of that, millennials rate general aspects of CSR as important (Barton et al. 2012; Leveson and Joiner 2014). In the context of sustainability however, environmental dimensions of CSR are perceived least important (ibid.). Accordingly, the majority of millennial job seekers express personal importance of sustainability, but almost all do not think that this would have any impact on their job search (Hanson-Rasmussen et al. 2014). Besides, non-environmental behaviour is negatively seen in severe cases and would then influence job decisions (ibid.). Millennials also recognise the increasing importance of sustainable matters once they enter the career path, since more job opportunities will evolve (ibid.).

2.7 Sustainability, Luxury, and Millennials

The millennial generation is characterised by a mindset and correlating values that are different from the ones of previous generations. Emerging technologies and the general trend of digitalisation are just some of the influences that shape the current zeitgeist. Thus, millennials influence industries, amongst other aspects, as both consumers and job seekers. Their awareness for environmental, social, and economic issues critically questions current business models. Besides, a different value distribution regarding luxury influences the relevance of such products to millennials.

On the assumption that millennials demand more transparency from luxury goods manufacturers and require labels to communicate that, *The Butterfly Mark* (Positive Luxury 2017) and the *Green Carpet Challenge* (Eco-Age 2017) exemplarily offer orientation points for sustainable luxury. Moreover, prominent figures use their influence to raise awareness for sustainability in all its facets. *Emma Watson*, the 26-year-old (millennial) British actress, for instance, besides her engagement for gender equality, promotes sustainable fashion using public appearance, for example at the annual *Met Gala* in 2016 where she wore a *Calvin Klein* dress made from recycled plastic bottles (Spedding 2016; Winston 2016). Additionally, in 2016, *Stella McCartney*, a British fashion designer and pioneer in sustainable fashion, launched a perfume that is designed especially for millennials and bottled in a packaging that minimises the environmental impact (Stella McCartney 2017). These examples show that sustainable luxury for millennials is a significant issue to investigate.

Consequently, a shift in the value proposition is expected. Whereas luxury goods manufacturers hitherto positioned themselves in dimensions of pomposity, profusion, and opulence, a paradigm change towards efficient, responsible, social, and smart luxury takes place. Millennials drive this change in the understanding of luxury. For example, millennial consumers can use the conspicuousness aspect of luxury goods to communicate their personal commitment to environmental protection by buying certain products that are obviously dedicated to benefit a specific cause (Cervellon and Shammas 2013). Consequently, consumers have the feeling that they have a positive impact on their surrounding by purchasing these products (Amatulli et al. 2017).

However, the question arises whether luxury consumers actually do consider sustainability as a selling point or rather have an aversion to it (Givhan 2015). Although CSR in particular is not yet fully implemented in the luxury goods industry (Höchli 2015), François-Henri Pinault, Kering's CEO, arguments that they act sustainably for economic reasons (Givhan 2015). The lack of communication of these efforts and transparency towards the consumer result partly from the fact that these firms do not see customer demand for CSR activities (Bhaduri and Ha-Brookshire 2011; Höchli 2015). On top of that, sustainability scandals worsen millennials' perceptions of luxury goods manufacturers. In 2013, it was uncovered that many of the products manufactured in the Tuscany (Italy), for brands such as Gucci and *Prada*, and labelled "Made in Italy" were actually fabricated by minimally paid Chinese immigrant workers under conditions like those in sweatshops ("Slaving in the lap of luxury" 2013). Moreover, the French luxury goods company *Hermès* that attracted negative attention in 2015 when *PETA* investigators revealed that crocodiles were raised and slain under unbearable conditions to provide the leather for Hermès' handbags ("Exposed: Crocodiles and alligators factory-farmed" 2017).

To summarise, the need for sustainable action increases throughout the economy. Hence, also luxury goods manufacturers are urged to act regarding CSR. However, their role as goods manufacturers and employers regarding sustainability frees up space for interpretation. In this respect, the millennial generation offers a trigger for change due to their unique mindset and a different emphasis on values than previous generations. Luxury firms must understand the desires and moral concepts of these millennial consumers in order to facilitate a sustainable future.

3   Research Design

This study aims at capturing a snapshot of the meaning of sustainability for luxury from the perspective of the millennial generation. Thus, focus group interviews are more appropriate than other data generation techniques. There is no need to gain deep personal-related insight into this topic, or to generate reliable data. The research conducted has a survey character that avoids socially desired answers and is not conceptualised to test hypotheses.

The present paper requires the crucial similarity of belonging to a certain generation amongst the focus group members. Therefore, the focus group was composed by the researcher rather than randomly. The shared background of experiences amongst the participants in order to facilitate the discussion is established through the same educational experience. Both male and female students of a Swiss business university form the focus group. To represent the millennial generation most appropriately, the six members were primarily selected by their year of birth ranging from 1980 to 1997. This even spread of cohorts is achieved by asking both bachelor and master students to participate in the focus group discussion. Because all participants are members of the same university, conflicts resulting from differences in status or class are non-existent. Furthermore, since no differences in gender opinions ought to be established, this focus group is mixed. Due to temporal resource constraints and the purpose of this research, two sessions with the same focus groups were held. These two meetings result from the fact that the same group of people was exposed to different circumstances. First, their argumentation originated from a customer perspective. Second, they had to argue from a job seeker approach. The two framings were chosen because the characteristics of millennials are a game changer in these aspects. Thus, a more extensive picture of opinions that are relevant for the industry can be drawn. Since these two perspectives are probable situations for the investigated millennials, they can argue in a comprehensible way (Snoy 2010). To ensure comparability of the framed statements, the very same people were interviewed. Therefore, two sessions were conducted within an interval of two weeks in order to provide the group members with enough time to detach from the content already established and to generate a fresh approach. The discussions were conducted and documented via audio recording at the *Behavioral Lab* of the University of St. Gallen. In general, a funnel approach in the questioning route was chosen to condense and concentrate a wide understanding of sustainability on specific situations and examples in relation with the luxury goods industry.

The evaluation is based on an unabridged transcript of the audio record of the two focus group sessions. Hereby, topics are clustered and brought into context with each other initially. Second, important statements are included in the discussion to corroborate conclusions based on the theory presented.

4   Findings

The following statements are based on unabridged transcripts of the focus group discussions' audio recordings. The assertions solely originate from the participants' opinions that were expressed during the two focus group sessions. In the first session, general definitions of and associations with the terms sustainability and luxury as well as sensible intersections from a consumer perspective are established. The second session, conducted with the same participants, presents a different framing. This time, millennials were asked to reason from a job seeker perspective on desirable employer qualities, general attractiveness of luxury goods manufacturers, and meaningful sustainability aspects within these firms.

4.1   Consumer Perspective

The first focus group discussion comprises (1) a general part on semantic, situational, and brand-related associations with the terms sustainability and luxury, not pretending a specific argumentation perspective, and (2) the assessment of luxury-related issues from the millennial consumer point of view. Thus, the participants were asked to note their definition of and associations with the term sustainability in general on adhesive notes to discuss them and bring them into context thereafter.

4.1.1   Sustainability

The group members attribute an ideological character to sustainability as one of the main issues. They refer to both positive and negative connotations. The positive ideological sustainability concern indicates a desirable status that should be pursued because of its positive impact. The negative ideological sustainability concern is associated with moralism and the fact that everything non-sustainable without further ado receives a negative image. In other words, when one decides to buy a sustainable product, for instance, she automatically excludes all other non-sustainable options ("It is a conscious purchase decision that you decide for a sustainable product and against the rest").

Thus, the first finding is that sustainability is perceived as a double-edged ideal conception.

Another key component of sustainability is the Triple-Bottom-Line with its ecological, economic, and social elements, whereas its meaning is especially biased by the ecological connotation in everyday linguistic usage ("[…] I do not think of balanced economic key figures but especially of organic vegetables"). Thus, corresponding associations arise, such as organic vegetables, environmentally friendly, and carbon dioxide footprint. The Triple-Bottom-Line's economic element is related to at last even after the social factor. Sustainability also relates to durability of products opposing disposable articles, generating long-term value, and enabling repair. In this respect, conscious consumption that is thought provoking in purchase decisions represents an additional key factor. It also comprises conserving (natural) resources in the production process and beyond as a premise for sustainability. This connects to a major topic of renewal that links to durability, renewable energy, waste separation, recycling, and environmental friendliness.

Thus, the second finding is that sustainability is associated only with the ecological dimension of the Triple-Bottom-Line.

The final major aspect of sustainability is future orientation, connected with a long-term time horizon and purpose as well as far-sighted consumption. In this context, generational responsibility plays a key role, too. The participants believe egocentric actions are short-sighted and responsibility for the current as well as future generations is crucial ("[...] and generational friendliness probably is just another expression for a forward-looking approach in the sense that we should not think of nothing but ourselves but also of other generations. I think that sustainability relates to that").

Thus, the third finding is that millennials regard to sustainability in a long-term, intergenerational way.

In summary, this means that millennials attribute both positive and negative aspects to sustainability, relate to it predominantly on the ecological level, and are aware that there must be a long-term responsibility for generations to come.

4.1.2 Luxury

The associations with luxury are of great variety. In general, the term luxury is interpreted in two different ways. On the one hand, it is associated with tangibles, such as jewellery, cars, art, yachts, represented by brands like *Rolex* and *Van Cleef and Arpels*. Here, it is also referred to product quality, precious materials, and intrinsic value. However, luxury products can also be used to symbolise status and are more generally observed as things that are superfluous, irrational, or at least not essential for everyday life. In the material context, luxury is linked to individualised products that connect to another major aspect, namely the individual understanding of luxury. This manifests itself particularly in the second category of luxury, the immaterial ("[...] luxury for me is the 'money cannot buy experience', so, something immaterial that gives me additional personal value and cannot be defined nor experienced through money"). Terms, such as travelling, time in general, spare time, vacation, and experiences, embody the intangible aspects of luxury. This aspect of the intangible and individual goes along with associations of individual exclusivity that relate to high-priced tangibles but also to inexpensive experiences.

Thus, the first finding is that millennial luxury is represented by mostly intangible and personalised experiences.

Other connections are drawn to the aspect of carefreeness not only in terms of political and financial security but also when it comes to purchase decision-making and final consumption. Also, education and political say, as part of freedom of choice, contribute to insouciance. Furthermore, luxury is also seen as something desirable, which is not reached yet, but will eventually be achieved when monetary preconditions are met. In this context, wealth is also seen as an enabler of luxury to finance desires, and of insouciance. Eventually, a certain intransigency increases the enjoyment of luxury (products), which does not necessarily imply waste of resources.

Thus, the second finding is that a cognitive carefreeness and education serve as enablers of luxury.

In summary, a clear tendency towards intangibles can be observed when assessing millennials' understanding of luxury. Individuality and personal fit also serve on the cognitive level as preconditions to enable luxury.

4.1.3   Combination of Sustainability and Luxury

A general commonality of sustainability and luxury is established in the two-tier nature of both of these terms. On the one hand, sustainability is rather distinctly defined by a holistic approach of economically, ecologically, and socially beneficial actions. On the other hand, sustainability is biased towards an ecological understanding in everyday usage. Similarly, luxury is highly individually defined, which results in a great variety of tangible and intangible associations. This describes the observed discrepancy between the terms sustainability and luxury ("[…] both of them are not unconditionally compatible, and one is preferred over and thus uncompromising towards the other, depending on the viewing angle"). Intransigency and consumption preferences exacerbate the conjunction of sustainability and luxury in many aspects. However, the future orientation of sustainability through long-term durability and value of luxury products, resulting from high manufacturing quality and fine materials, establishes a precondition for sustainable alignment.

Thus, the first finding is that sustainability and luxury share a minimal basic conformity.

The prevalent basis for connecting sustainability and luxury are products, the way they are manufactured, the sourcing of their raw materials, and their way of usage. In this context, generational awareness and responsibility connect to luxury products in the way that these objects obtain individual emotional values depending on their origin, for example as a gift from a closely related person, and personal experience with these products. On the other hand, intangible luxuries do not per se relate to sustainability because there is no resource consumption. However, also the way one spends her free time, for instance, could harm the environment. Besides, ideational values as part of intangible luxury connect to the social part of the Triple-Bottom-Line and thus to ethical ways of sourcing raw materials, for instance.

Thus, the second finding is that luxury products offer the most obvious connection to sustainability.

Although there is the wish for a holistic sustainability approach that does not reduce efforts to single processes in the supply chain, some actions, such as autonomously generating electricity from renewable resources, are perceived as out of scope of luxury firms' core competencies. Because the participants think that they as customers do neither have insight into nor influence on steps of the supply chain, they see their task in assessing whether the product itself is sustainable and how they can use it sustainably instead of investigating the degree of sustainability across the supply chain. Hence, the participants wish for transparency in luxury companies' supply chains as well as honest communication of sustainability efforts to facilitate purchase decision-making ("It is two different things: what is communicated and what actually happens, and whether it is communicated because the company wants to be sustainable or because they pursue marketing efforts").

In comparison with the food industry, for example, an increasing demand for transparency from the customer side and an increasing supply from the manufacturer side is reckoned. In this example, transparency has become a right to exist for food producers. Sustainability hallmarks, which are custom in this industry, are a suitable tool for only some millennials to assess the source of a luxury product and give guidance as well. Others think hallmarks would decrease the perceived value of luxury products and yet remain intransparent regarding actual compliance with the underlying criteria. They argue that hallmarks are not suitable for a positive distinguishing feature ("I would find it dumb if eco-seals or the like were placed on such products"). Instead, the fact of luxury itself, the trust in a luxury brand as a hallmark, should be used. As one focus group participant expressed: "I do not want to have to think about whether luxury products are sustainable." Hence, basic sustainability, resulting from general product qualities, is assumed and desired.

Thus, the third finding is that millennials demand transparent sustainability actions to some extent. Since basic sustainability of luxury products is assumed, eco-labels embody a disadvantageous way to communicate sustainability.

Luxury companies that do communicate sustainability are critically observed as well ("[…] is everything sold as sustainable really sustainable?"). It is questioned whether these firms want to be sustainable and thus communicate that they are or they use the publicity only for reputation purposes. That this issue is rather a question of branding than of sustainability is attributed to classical luxury labels.

Thus, the fourth finding is that millennials still question transparent sustainability communication of luxury goods manufacturers regarding its authenticity.

While some participants argue that the origin of resources is less questioned with luxury goods, and unconcern in usage is practiced, others pay special attention towards the formation process. For example, in the case of rather small luxury goods, the importance of social sustainability prevails, since the impact of non-sustainable resources is limited. The necessity of ethical sourcing efforts is primarily associated with third-world countries because the farther you go back in the supply chain, the more likely it is that sustainability efforts are necessary and sensible. For instance, if sourcing of diamonds is restricted to certain regions, it should be communicated that mining is executed under ethical and socially sustainable standards. Additionally, one must differentiate between luxury goods, such as watches, and luxurious consumables, such as wine. In this context, the participants care less about sustainable luxury goods than about sustainably produced luxury consumables.

Thus, the fifth finding is that social and ecological sustainability primarily matter for millennials in the sourcing process of luxury goods.

Furthermore, it is perceived that the relevance of sustainable consumption depends on age due to an altering assessment. This also relates to brand consciousness and status consumption that decrease in relevance with increasing age, also due to rising disposable income. Moreover, individuality of luxury products rules out the altruistic understanding of sustainability, which is important in general, but the personal fit of nice products is more significant ("I would not say that sustainability is unimportant to me [...] if I like something better [...] I buy it" "Yes, I would do the same").

Thus, the sixth finding is that personal fit and age of millennials significantly influence the relevance of sustainability in luxury for them.

On top of that, the participants note current trends of sustainability, also about innovation, for instance *Tesla* or the *Toyota Prius*, in their environment. In this context, fringe groups seem to initiate sustainability trends, for example in the car industry. Although, these groups comprise mostly people, who have the financial ability and can afford to purchase such cars in addition to already existing car pools. It is seen as an act of public demonstration in correlation with positive societal perception. In fact, respective peers and the surrounding society influence one's purchase decision process. Hence, sustainable products can be a symbol of status because one belongs to a certain group and sets an example. A Prada handbag, for instance, is then purchased not because it is luxurious but because it is sustainably manufactured. But also on the personal usage level of luxury goods, the factor of sustainability is suited to silence one's conscience and to justify purchases of luxurious brands towards peers and society.

Thus, the seventh finding is that sustainability in luxury can be used for green status consumption, as a trigger for innovation, and as a justification of personal luxury consumption.

Education, as another part of sustainability in a societal context, is perceived as a luxurious premise to gain awareness of sustainability and luxury correlations in the first place. On top of that, personal as well as general financial resources enable environmental friendly actions and allow to fund ecologically reasonable actions that are not yet economically sustainable. Hence, wealth creates room for manoeuvre and offers selection options. It is an enabler of beneficial actions, such as in the example of the *Bill and Melinda Gates Foundation*.

Thus, the eighth finding is that educational and financial resources enhance the awareness and impact of sustainability in luxury.

Concerning negative examples of luxury firms that run non-sustainable processes, participants claim that they would decide against the brand or the product because they have the choice not to buy and look for alternatives. Moreover, the need for products would be questioned if these were produced in a scandalous way, since the price premium charged for such luxury products would not reflect the intended quality and aspiration anymore. Generally, scandals would be more shocking with luxury products than with regular products due to the supposition of a basic degree of sustainability. However, personal reactions are also made conditional on society's reception and individual moral pain thresholds. It is also established that such non-sustainable behaviour would not have any influence on intangible luxury.

Thus, the ninth finding is that sustainability scandals are observed as more severe within luxury than in other industries and thus lead millennials to refrain from buying evident non-sustainable luxury goods.

Eventually, intersections of sustainability and luxury are limited to a product base and do not entirely represent the participants' associations with luxury ("To be honest, I do not necessarily associate luxury goods with sustainability. That would rather be jute clothes from a mountain village in Spain"). It is depicted that sustainability is a side issue within the luxury goods industry. There is no unique luxury brand that is associated with sustainability, also partly because there is no expectation for sustainable luxury firms in the ecological sense. As one participant explained: "if you want to buy a sustainable product, you do not buy a Louis Vuitton handbag". While some participants stated that the disposition to buy luxury goods increases when they are produced transparently sustainably, others think that sustainability does not add a surplus to luxury goods, it rather limits their appeal and enforces compromise.

Thus, the tenth finding is that sustainability is currently not identified within the luxury goods industry, and the benefit of its implementation remains questionable.

In summary, the intersection of sustainability and luxury is mainly limited to the sourcing of luxury products' materials, whereas transparency about luxury goods manufacturers' operations is desired, yet would be critically seen. Moreover, the importance of sustainability in luxury depends on each millennial individually, whereas sustainability can be used to trigger advantageous development within the luxury industry. In this regard, financial resources are key. Although millennials perceive sustainability scandals within luxury as more severe than elsewhere, they cannot observe sustainability efforts within the industry.

These findings, their interconnection, and overarching topics as well as contradictions in millennials' statements are clearly comprised in Fig. 1. To conclude, both sustainability and luxury are defined in fuzzy and subjective ways. Their intersections are very limited and concentrate under the topics of transparency, holism, and value chain reference. Contradictions in millennials' statements evolve across the case-related associations with sustainability scandals and the consumer perspective. The conclusion evolves that sustainability and luxury are two very different animals.

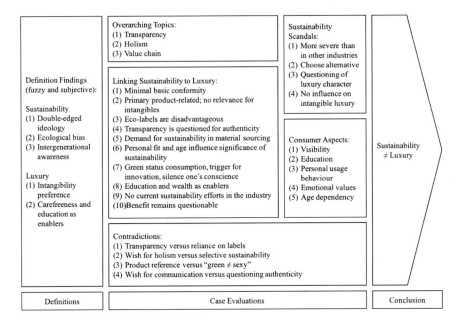

Fig. 1 Key findings from consumer perspective (own illustration)

4.2 Job Seeker Perspective

In the second focus group discussion, the very same participants were invited to talk about the topics of sustainability and luxury from the perspective of a job seeker in order to establish comparability of the answering patterns. First, they were asked to name attractive potential companies as employers, possible industries of occupation, and desirable employer qualities. Second, the rationale why luxury goods manufacturers are attractive employers was established. Third, participants were asked to identify essential aspects in luxury goods manufacturers' value chains that should be sustainable, such that these firms are (more) attractive potential employers. Last, it was assessed how negative examples of non-sustainable luxury goods manufacturers would influence participants' decision for a luxury employer regarding (1) future professional development and (2) the standing in their social environment.

4.2.1 Desirable Employer Qualities

Initially, it was identified that several aspects interact and create an overall balance, but also force trade-off situations. Possible employers are: *Universal Music,* Kering, *Infosys, EY Parthenon, Mars, Beiersdorf,* LVMH, Richemont, and *Montblanc.* The respective industries emerged: music management, luxury goods, consulting, consumer goods, private equity funds, and marketing. It is notable that younger millennial participants are not sure about specific employer names or even industries.

Then, regarding important characteristics of a desirable job opportunity, participants seek for a comforting working environment, consisting of nice facilities, companionable co-workers, and a culture that is open for new ideas, creativity, as well as change and disruption:

> "[…] that the firm and the working environment are open to new ideas and creativity, that you can actively participate"; "I definitely want that an openness to change, an openness to disruption. For me that is recipe for the firm to evolve, for the working environment to evolve […] I can suggest new ideas and thus an open attitude is lived within the company"

In this context, participants emphasise the importance that they can have an impact through their work ("[…] that you have the feeling to have an impact and are not just a small wheel in the company"). They do not want to have contact to managerial staff and take opportunities to develop within and alongside their professional path. Besides, a diversified and challenging field of duties is desired that keeps the participants from being bored. This shall be enabled by flexibility regarding working hours and different job locations as well as an equilibrated work-life balance. Additionally, adequate compensation is perceived as a prerequisite rather than a unique feature; intangible job qualities predominate ("If I am after better payment, I would probably look for a different industry […] and thus would not consider what I had as predilection and wanted to do intrinsically. This would rather fulfil me than a higher salary").

Thus, the first finding is that millennial job seekers look for adaptability of their work environment and the firm as well as opportunities to have influence in their jobs.

4.2.2 Employer Attractiveness of Luxury Goods Manufacturers

Because these luxury products enrich people's lives, trigger special feelings, deliver an intangible surplus, are beautifully made, offer personal identification, are communicated in an exceptional way, and tell unique stories, the participants create interest in the luxury industry ("[…] because luxury goods manufacturers can make people very happy"; "[…] in many regards, [luxury goods] add a much greater immaterial value for the consumer"). Moreover, fascination for luxury goods firms results from the perceived enthusiasm of employees, the lifestyle and prestige of the industry, high customer appreciation, opportunities to work on international level, the financial prosperity that gives leeway to unfold creativity and indicates higher wages, and the strategic challenges, such as the conflict of reinvention and brand fidelity as well as the leverage of synergies within luxury conglomerates. In addition, working for luxury employers is seen as obtaining a free ride for future occupations.

On the contrary, there are also critical statements that formulate why luxury goods manufacturers could not be attractive employers. As perceived by the participants, some luxury goods firms live in their own bubble, are resistant to progress and change, poise in tradition, and rest on their laurels. Besides, a certain level of arrogance is perceived that results from cultural issues within the industry. In context with the previously stated desire for having an impact, one participant expresses that she thinks she would have less impact on society with limited intergenerational passing on than in other industries ("There are so many other things [than the luxury industry] [...] where I have more impact on society or the world [...] where I pass something on to the following generation [...]"). Another focus group member sees this lack of sustainability as a motivational factor to get involved and initiate change.

Thus, the second finding is that millennials can imagine to work for luxury goods manufacturers but also fear that corporate inertia and industry pride could limit their impact.

4.2.3 Combination of Sustainability and Luxury

First, an economically sustainable future—not relying on an outdated business model—is identified as the premise to act sustainably in other aspects:

> "I do not care if they publish a carbon dioxide report. I think it should be the case that I have the feeling that the company has good prospects and does not become totally obsessed by an outdated business model. As far as this is given, it is not that important to me how their energy footprint looks like"

However, only a holistic sustainability approach, which cannot be perceived yet, that includes all aspects of the Triple-Bottom-Line ensures a luxury company's future. The impact of sustainability efforts of luxury goods manufacturers on job seekers is discussed in a variety of aspects. While some argue that these efforts would not have any impact on the decision for or against an employer because job seekers are glad to get any job offer, others argue that it depends on the individual predilection for sustainability. In this respect, participants observe a prevailing economic motivation behind (luxury) firms' sustainability endeavours. Some focus group members can neglect the motivation behind such actions as long as the outcome is beneficial, while others critically question sustainability projects for authenticity.

On top of that, the relevance of sustainability actions, once one is working in a luxury company, depends on their occupation and personal involvement in the respective actions. In this context, both relevance of and impact on sustainability efforts within luxury firms depend on one's career stage. While early on, minimal impact in the company and limited personal relevance, related to product identification, are prevalent, later career stages offer impact opportunities and increase personal relevance of the sustainability issue ("I just think that sustainability influences my daily work less as a job entrant, and that strength becomes prominent when I climb up the career ladder and then recognise I could change something").

Thus, the third finding is that millennials perceive an increasing relevance of sustainability in the employer selection process with advancing career, based on existing personal inclination.

The relevance for future professional development initially originates from the respective career stage ("[…] depending on which stage you are, it [sustainability] has differing significance"). Because first-time job seekers rather take opportunities to fill gaps in their curriculum vitae and have little impact, it does not matter as much. However, one could also purposefully look for sustainability challenges with luxury goods manufacturers to get engaged ("I think it would be interesting to say that I go into the luxury goods industry and I want to have an impact"; "[…] maybe I can have an impact to change things"). In a later stadium of one's professional life, these scandals become more relevant because long-term occupations are pursued ("If you then actually look for a permanent occupation in the long-term, I think, this [employer differentiation based on sustainability] is clearly more relevant"). Also, then you rather have more job options to choose from and are not reliant on every opportunity there is. In fact, you could quit the current profession and look for other occupations if your ethical standards do not comply with company practice. Thus, it also depends on your personal career goals, whereas the participants do not see relevance in the luxury goods industry ("I think it depends on where you want to go. So, if you aim for a job in environmental protection, it is not advantageous that you worked with *Nestlé* before. I do not think this is relevant in the luxury industry"). In other words, a former job at a scandalous firm will not hinder you from going into the luxury industry ("[…] the majority of us would turn their back on such a scandal").

On top of that, the relevance depends on the scandal's extent and duration. If there is only a minor sustainability issue in a luxury firm, no impact on the professional future is expected ("[…] in no way would it be a disqualifier to apply at this firm, why would it be?"). Likewise, personal involvement in the type of scandal as well as in the causation of the scandal matter. Additionally, participants would refrain from applying to luxury goods manufacturers where non-sustainable behaviour originates from corporate culture issues ("I think it is exactly that: If it is correlated with the corporate culture and if you have the feeling it is a fundamental problem […] then it would be something negative for me and it would upset me"). In this case, the issue of transparency plays a crucial role for the participants' decision-making. For example, if it is publicly known that a certain luxury goods company has a sustainability scandal due to poor corporate culture, focus group members would not consider this firm as a potential employer.

Regarding the participants' standing in their respective milieu, no concerns are expressed if they were employed with a non-sustainable luxury goods manufacturer. Participants do not expect major criticism from their like-minded peers. As long as there is no direct connection between the millennials' actions and the respective scandal, they know how to justify their employer decision towards others.

Thus, the fourth finding is that millennials make environmental scandals conditional on their current career stage, the scandal's extent and duration, their personal involvement, and the corporation's scandal culture. Millennials exclude any reputation loss in their social environment.

A summary of the findings according to major topic areas can be depicted in Fig. 2. To conclude, millennials ascertain a range of employer qualities that are important for them. In this respect, corporate culture qualities are emphasised. The aspect of personal fit links millennials to luxury goods manufacturers as employers. The reconciliation of sustainability in luxury from a job seeker point of view peaks in the dependence on impact. This is the central issue that decides about the relevance millennials see in sustainable luxury goods manufacturers.

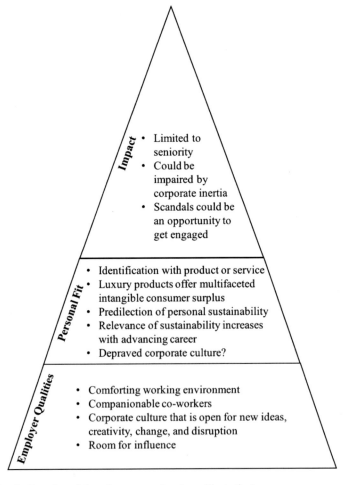

Fig. 2 Key findings from job seeker perspective (own illustration)

5 Discussion

As established above, millennials embody a generation of critical thinkers and create a diversified image of opinions. The focus group participants of this study confirm with the stated definition of millennials. Hence, it is of no surprise that their perceptions and interpretations are diverse on individual level and thus form a multi-faceted overall image.

It can be stated that the millennial participants have a general awareness of ecological importance. The Triple-Bottom-Line is identified as a basic approach to sustainability, and the bias towards mostly ecological associations is observed by the focus group as well. Despite their knowledge about the social and economic dimensions, these millennials stick to the biased view throughout the discussion and express mainly ecological aspects. It is notable that these business students do not associate, for example, economic long-term stability directly with sustainability in a broader context. Far reaching consequences of unsustainable behaviour are only partly addressed in the form of intergenerational awareness and general long-term alignment. Associations and contextualisation on the macro level did not evolve. Thus, holistic knowledgeability about sustainability cannot be observed. This does not indicate that the participants are not knowledgeable in the respective area. However, it leads to the assumption that coherencies on a large scale are not that important to the participants. Moreover, the millennials identify the possibility to fund sustainable actions that are not yet profitable.

With respect to sustainable products, durability, reparability, and long-term usage are the key words. Hence, these millennials rather see it as their duty to make a product sustainable by preserving it over a long period of time and giving it to the next generation. This is corroborated by conscious consumption, which coerces the participants not to buy cheaper, more breakable products.

In their opinion, sustainability also comprises a virtuous dimension, which labels unsustainable actions as evil. Therefore, they also argue that (personal) non-sustainable behaviour would not be severely judged by their like-minded peers. Hence, it can be assumed that sustainability as a concept does not play a key role.

The aspect of individuality and personal interpretation can be identified in the definition of luxury as well. Expectedly, luxury is not defined in a single and unambiguous way. Instead, a range of mostly intangible experiences and product-related associations are elaborated. Exclusive experiences, which do not have to be expensive, build the core of these millennials' luxury definition. However, the conclusion immediately suggests itself that the interviewed focus group, compared to millennials from other cultures, is rather able to consume luxury products and thus thinks of luxury goods in a more convenient way.

Moreover, a good understanding of the vital characteristics of luxury goods can be observed. A major role, however, plays the durability and timelessness of such products. These elements are preserved through pass on to the following generation, for instance. Again, product-related emotional values and experiences supersede the material value. In other words, purchase intentions result from personal identification rather than representative reasons. Thus, inclination to buy luxury products seems to be mitigated because personal identification, emotional connection, and intangible luxury prevail.

Literature claims that millennials simultaneously desire to protect the environment and seek to enjoy the products they consume. From the findings of the first focus group session, this behaviour is hardly enacted because beauty of and desire for certain products outweigh the need to act for the greater good, unless major scandals, for instance, are perceived to be severe by peers and society as well. Additionally, an age-related importance of conspicuous consumption is sensed: Younger participants rather gain social approval by purchasing certain brands and products, while elder millennials do not value alignment with peer group that much.

The combination of sustainability and luxury raises diverse and non-compliant understandings. Contrary to the often-claimed characteristics luxury goods possess to enable sustainability, participants perceive a very limited number of intersections between luxury and sustainability. On the one hand, luxury is associated with durability and intergenerational value retention. On the other hand, the uncompromising character of such goods prevails. While some want to rely on their conviction that luxury products are sustainable, others are convinced that luxury firms do not act sustainably at all and thus see opportunities for these companies to put more effort in sustainable and innovative development also because green innovation is appealing to them. Moreover, due to mainly ecological associations, openly communicated sustainability in luxury companies even represents a negative differentiator in some participants' view. This leads to differing importance of sustainability in the image of luxury firms. The millennial consumers are not aware of sustainable luxury firms, although many of these companies are highly active in CSR and communicate their activities quite openly.

Besides, some millennial participants imagine sustainable luxury not only as a justification towards peers but also to ease one's mind about the products pricing. They want to shuffle out of responsibility. In fact, this contradicts their often-claimed desire to have an impact. Moreover, the participants are not aware of current unethical behaviour in the luxury industry. Yet, if significant non-sustainable issues arise in production of luxury goods, claims towards such products are no longer met, and alternatives are examined.

The established congruencies and deviations show that the ideology of sustainability in luxury has not yet caught on in millennials' minds. This generation only partly takes sustainability efforts into account when purchasing luxury goods. Mostly, personal experience related, intangible factors and the affiliation to the product itself drives consumer behaviour. This may result from the fact that the interviewed millennials not yet possess the financial resources and thus care more about the personal fit of a respective product than its (non-)sustainable way of production.

However, the issue of transparency is a major aspect where luxury firms can build on to gain trust of millennial customers. This means that communication of sustainable product-related engagement is desired by millennial customers, whereas operational sustainability does not primarily have relevance. The participants also correctly identified that luxury goods manufacturers deliver an emotional, intangible value that goes beyond the mere material equivalent by telling unique stories. This factor could be used to promote sustainable products in an appealing way that distinctively uses the brand's heritage without appearing overly green.

Participants' expressions regarding desirable employer qualities are congruent with data observed in research so far. The focus group members demand mostly individual elements, such as work-life balance, challenging tasks, and a comforting work environment, whereas a trade-off between these elements is stressed. They want to work for companies they have passion for and can identify with rather than those which are fashionable. In addition to current research, the company's openness to change is emphasised throughout the discussion, whereas luxury firms could be associated with inertia. This factor offers a notable characteristic. Millennials, who live through some major disruptions, such as digitalisation, seek for adaptability to change to avoid stagnation. This motivation should be used by bringing environmental issues to millennials' attention and implement sustainability measures in the value chains of firms they want to work for.

Comparable to the first session, the importance of sustainable action in general is identified and a basic sense for sustainability is demanded. Again, transparency about the real motivation behind a company's sustainability efforts is desired because primarily economic justification makes millennials leery. These millennials do not observe sustainability to be relevant for their current job seeker situation because they underestimate their impact as career entrants although they emphasise that they want to have an impact wherever possible: on end consumers, the society, and the world in general. Yet, they imagine that the importance of sustainability will increase with proceeding career due to growing personal influence of their work that enables additional opportunities. In this context, a shift in decision-making relevant factors from "I will take almost any job offer" to "I want an occupation that exactly meets my needs" will take place.

Overall, the interviewed millennials argument in a reflective and differentiating way. They consider both present status and future possibilities of their actions. It is identified that once they have the possibility—financially as consumers and for career reasons as long-term employees—the importance of sustainability in luxury will increase for them personally. However, this leads to the question why they do not use all current possibilities to act in a sustainable manner. Even though their current capabilities are limited to some extent, these millennials could still have a more significant impact as consumers and job seekers. For example, environmentally conscious millennial job seekers could influence sustainable luxury by founding their own companies combining luxury and sustainability in an innovative way.

6 Conclusion

This study generates a snapshot of millennial opinions on the issue of sustainability in the luxury goods industry. It thus adds to already existing research and helps filling the gap of millennials' impression of sustainable luxury. This focus group research establishes that millennials, despite their general awareness of sustainability's importance, have limited motivation to use current impact possibilities to make a change. Moreover, they do not favour sustainability of luxury firms because they prefer intangible luxury where the sustainability lies in their hands depending on personal usage behaviour. Thus, sustainability with luxury goods is reduced to durability resulting mainly from millennials' way of use. Similarly, basic sustainability aspects of luxury goods manufacturers are expected by millennial customers and job seekers. Having said this, millennial job seekers do not see sustainability as a disqualifier when differentiating between potential employers. Eventually, their growing personal influence makes millennials think that they will increasingly impact such companies both as customers and as employees regarding sustainability efforts.

6.1 Implications for Practice

The discrepancy between sustainability efforts communicated by luxury goods manufacturers and millennials' perception of these CSR activities leads to a two-tier implication: On the one hand, it could indicate that millennial consumers have no inclination to question whether luxury products are sustainably manufactured and thus do not inform themselves. On the other hand, the lack of awareness could result from insufficient promotion of sustainable efforts from the manufacturers' side. In both cases, a mismatch between consumer and producer side can be established.

This can advise companies to particularly emphasise non-ecological parts of the Triple-Bottom-Line in order to sensitise millennials for the economic and social qualities of their products.

Based on the study's findings, our key implications for luxury goods manufacturers in practice are:

(1) Utilise financial resources to fund sustainability projects that are not yet economically sustainable and intensely communicate efforts and respective achievements.
(2) Use product customisation, brand-related communication, and unique events to address millennials on a personal level where they can identify with you as a sustainable luxury goods manufacturer.
(3) Do not actively communicate your sustainable products as "green luxury", but emphasise sustainability as a positive side effect that makes these goods even more desirable.

(4) Transparently disclose the operations along your supply chain and be open and honest about the actual motivation behind sustainability efforts to help millennial consumers to comprehend the origin of desired luxury goods and assess you as a potential employer.

(5) Emphasise that your company as an integrative employer is open to change and disruption.

(6) Promote low level sustainability projects to signalise job seekers that career entrants have an impact regarding sustainability in your firm.

(7) Creatively surprise and delight millennial consumers and job seekers by connecting designers to sustainability.

(8) Establish and stand for a sustainability purpose that is meaningful in millennials' perception.

6.2 Implications for Research

Although the findings of this study are not generalisable to an entire generation, we have established that millennials' opinion about the terms sustainability and luxury as well as their intersection from a consumer and a job seeker point of view are not only multi-faceted but also include some contradictions. Therefore, it is of utmost relevance that these findings trigger further investigations regarding the relevance of sustainability in luxury from millennials' point of view. Here, we kindly ask our fellow researchers to enrich the generated picture and gain validation beyond its scope also considering the following outlook.

The questions raise, which measures luxury firms need to implement to facilitate millennials' exertion of influence in particular, and what can be done in order to sensitise following generations for the necessity of environmental-friendly action, such that young people act according to their individual capabilities. On top, it would be interesting to investigate in future research how millennials from other cultures and societies assess the topic of sustainable luxury. Thereby, the present study could be replicated and extended by quantitative investigations on a larger, international scale.

References

Ahrend K-M (2016) Geschäftsmodell Nachhaltigkeit: Ökologische und soziale Innovationen als unternehmerische Chance. Springer Gabler, Berlin

Amatulli C, De Angelis M, Costabile M, Guido G (2017) Sustainable luxury brands: evidence from research and implications for managers. Palgrave MacMillan, London

Barton C, Fromm J, Egan C (2012) The millennial consumer: debunking stereotypes. https://www.bcg.com/documents/file103894.pdf. Accessed 2 Feb 2017

Barton C, Koslow L, Beauchamp C (2014) How millennials are changing the face of marketing forever. Bcg. perspectives. https://www.bcgperspectives.com/content/articles/marketing_center_consumer_customer_insight_how_millennials_changing_marketing _forever/. Accessed 2 Feb 2017

Bastien V, Kapferer J-N (2013) More on luxury anti-laws of marketing. In: Wiedmann K-P, Hennigs N (eds) Luxury marketing: a challenge for theory and practice. Springer Gabler, Wiesbaden, pp 19–34

Bergh C (2017) Demanding millennials buy into brands that authentically embody their values: how a timeless classic illustrates the millennials' need for substance. https://www.atkearney.com/america250/demanding-millennials-buy-into-brands-that-authentically-embody-their-values. Accessed 2 Feb 2017

Bhaduri G, Ha-Brookshire JE (2011) Do transparent business practices pay? exploration of transparency and consumer purchase intention. Clothing Text Res J 29(2):135–149. doi:10.1177/0887302X11407910

Bian Q, Forsythe S (2012) Purchase intention for luxury brands: a cross cultural comparison. J Bus Res 69:1443–1451. doi:10.1016/j.jbusres.2011.10.010

Bolton RN, Parasuraman A, Hoefnagels A, Migchels N, Kabadayi S, Gruber T, Loureiro YK, Solnet D (2013) Understanding generation Y and their use of social media: a review and research agenda. J Serv Manage 24(3):245–267. doi:10.1108/09564231311326987

Buerke A (2016) Nachhaltigkeit und Consumer Confusion am Point of Sale: Eine Untersuchung zum Kauf nachhaltiger Produkte im Lebensmitteleinzelhandel. Springer Gabler, Wiesbaden

Bulthuis QWW (2015) Millennial luxury: an analysis into the generational cohort's values and perceptions of luxury. Master's thesis, Universitaet St Gallen's Catalog (Accession No. stgal.000860055)

Carcano L (2013) Strategic management and sustainability in luxury companies: the IWC case. J Corp Citizsh 52:36–54

Cervellon M-C, Shammas L (2013) The value of sustainable luxury in mature markets: A customer-based approach. J Corp Citizsh 52:90–101

Corporate Knights (2015, January 21) 2015 global 100 methodology. http://www.corporateknights.com/reports/2015-global-100/methodology/. Accessed 2 Feb 2017

Corporate Knights (2016, January 20) 2016 global 100 results. http://www.corporateknights.com/reports/2016-global-100/2016-global-100-results-14533333/ Accessed 2 February 2017

Cowdrey D (2007, November 29) New WWF-UK report ranks luxury companies and advises celebrities not to promote dirty brands. http://www.wwf.org.uk/updates/new-wwf-uk-report-ranks-luxury-companies-and-advises-celebrities-not-to-promote-dirty-brands. Accessed 2 Feb 2017

Doval J, Batra GS (2013) Green buzz in luxury brands. Rev Manage 3(3/4):5–14

Eastman JK, Liu J (2012) The impact of generational cohorts on status consumption: an exploratory look at generational cohort and demographics on status consumption. J Consum Mark 29(2):93–102. doi:10.1108/07363761211206348

Eco-Age (2017) The green carpet challenge. http://eco-age.com/gcc-brandmark-brands/. Accessed 2 Feb 2017

Exposed: Crocodiles and alligators factory-farmed for Hermès "luxury" goods (2017) PETA. http://investigations.peta.org/crocodile-alligator-slaughter-hermes/#lightbox[group-2290]/1/. Accessed 2 Feb 2017

Givhan R (2015, December 8) Luxury fashion brands are going green. But why are they keeping it a secret? The Washington Post. https://www.washingtonpost.com/lifestyle/style/luxury-fashion-br...8-8c8a-11e5-acff-673ae92ddd2b_story.html?utm_term=.d6b1874ebd43. Accessed 7 Mar 2017

Guillot-Soulez C, Soulez S (2014) On the heterogeneity of generation Y job preferences. Empl Relat 36(4):319–332. doi:10.1108/ER-07-2013-0073

Gurtner S, Soyez K (2016) How to catch the generation Y: identifying consumers of ecological innovations among youngsters. Technol Forecast Soc Change 106:101–107. doi:10.1016/j.techfore.2016.02.015

Hanson-Rasmussen N, Lauver K, Lester S (2014) Business student perceptions of environmental sustainability: examining the job search implications. J Manag Issues 26(2):174–193

Hershatter A, Epstein M (2010) Millennials and the world of work: an organization and management perspective. J Bus Psychol 25:211–223. doi:10.1007/s10869-010-9160-y

Hill J, Lee H-H (2012) Young generation Y consumers' perceptions of sustainability in the apparel industry. J Fashion Market Manage: Int J 16(4):477–491. doi:10.1108/13612021211265863

Höchli M (2015) CSR in the luxury goods industry. Master's thesis, Universitaet St Gallen's Catalog (Accession No. stgal.000849477)

Hudders L (2012) Why the devil wears Prada: consumers' purchase motives for luxuries. J Brand Manage 19:609–622. doi:10.1057/bm.2012.9

Hwang J, Griffiths MA (2017) Share more, drive less: millennials value perception and behavioral intent in using collaborative consumption services. J Consum Market 34(2):132–146. doi:10.1108/JCM-10-2015-1560

Japan earthquake: Explosion at Fukushima nuclear plant (2011, March 12) BBC News. http://www.bbc.com/news/world-asia-pacific-12720219. Accessed 2 Feb 2017

Kapferer J-N (1998) Why are we seduced by luxury brands? J Brand Manage 6(1):44–49

Kapferer J-N, Laurent G (2015) Where do consumers think luxury begins? a study of perceived minimum price for 21 luxury goods in 7 countries. J Bus Res 69:332–340. doi:10.1016/j.jbusres.2015.08.005

Kering (2017a) Environmental pandl. http://www.kering.com/en/sustainability/epl. Accessed 28 Feb 2017

Kering (2017b) 2025 sustainability strategy: crafting tomorrow's luxury. http://www.kering.com/sites/default/files/kering_2025_sustainability_strategy_-_press_kit_0.pdf. Accessed 28 Feb 2017

Kirchhof A-K, Nickel O (2014) Marken nachhaltig erfolgreich führen. In: CSR und Brand Management. Springer, Berlin

Leveson L, Joiner TA (2014) Exploring corporate social responsibility values of millennial job-seeking students. Education + Training 56(1):21–34. doi:10.1108/ET-11-2012-0121

LVMH (2017) LVMH commitments. https://www.lvmh.com/group/lvmh-commitments/. Accessed 2 Feb 2017

Mahler D (2017) Don't give up on millennials: boldly offer what they personally value. https://www.atkearney.com/america250/don-t-give-up-on-millennials. Accessed 2 Feb 2017

Martínez P, Pérez A, Rodríguez del Bosque I (2014) CSR influence on hotel brand image and loyalty. Acad Rev Latinoam de Administración 27(2):267–283. doi:10.1108/ARLA-12-2013-0190

Meadows DH, Meadows DL, Randers J, Behrens WW (1972) The limits to growth—a report for the club of Rome's project on the predicament of mankind. Universe Books, New York

Millennials coming of age (2017) http://www.goldmansachs.com/our-thinking/pages/millennials/. Accessed 2 Feb 2017

Millennials outnumber baby boomers and are far more diverse (2015, June 25). http://www.census.gov/newsroom/press-releases/2015/cb15-113.html. Accessed 2 Feb 2017

Montesa F, Rohrbeck R (2014) Luxury organizations and responsibility: a toolbox. In: Berghaus B, Müller-Stewens G, Reinecke S (eds) The management of luxury: a practitioner's handbook. Kogan Page Limited, London, pp 409–421

Moroz M, Polkowski Z (2016) The last mile issue and urban logistics: choosing parcel machines in the context of the ecological attitudes of the Y generation consumers purchasing online. Transp Res Procedia 16:378–393. doi:10.1016/j.trpro.2016.11.036

Ng ESW, Schweitzer L, Lyons ST (2010) New generation, great expectations: A field study of the millennial generation. J Bus Psychol 25:281–292. doi:10.1007/sl0869-010-9159-4

Petersen T (2012) Nachhaltiges Wirtschaften – ganzheitliche Strategien und Prinzipien. https://www.bertelsmann-stiftung.de/fileadmin/files/BSt/Publikationen/GrauePublikationen/Policy-Brief-Nachhaltiges-Wirtschaften-de_NW_03_2012.pdf. Accessed 7 Feb 2017

Pomarici E, Vecchio R (2014) Millennial generation attitudes to sustainable wine: an exploratory study on Italian consumers. J Cleaner Prod 66:537–545. doi:10.1016/j.jclepro.2013.10.058

Popovici V, Muhcina S (2015) An overview of millennials' coming of age: the emergence of generation Y and its underlying and consequential socio-economic aspects. Ovidius University Annals: Economic Science Series 15(1).341–346

Positive Luxury (2017) What we do. https://www.positiveluxury.com/about/us/. Accessed 2 Feb 2017

Pufé I (2014) Nachhaltigkeit. UVK Verlagsgesellschaft mbH, Konstanz

Richemont (2016) Corporate social responsibility 2016. https://www.richemont.com/images/csr/2016/csr_report_2016.pdf. Accessed 2 Feb 2017

Schade M, Hegner S, Horstmann F, Brinkmann N (2016) The impact of attitude functions on luxury brand consumption: an age-based group comparison. J Bus Res 69:314–322. doi:10.1016/j.jbusres.2015.08.003

Seo Y, Buchanan-Oliver M (2015) Luxury branding: the industry, trends, and future conceptualisations. Asia Pac J Mark Logistics 27(1):82–98. doi:10.1108/APJML-10-2014-0148

Shukla P, Banerjee M, Sing J (2016) Customer commitment to luxury brands: antecedents and consequences. J Bus Res 69:323–331. doi:10.1016/j.jbusres.2015.08.004

Slaving in the lap of luxury: "Made in Italy" is not what it used to be (2013, March 1) The Fashion Law. http://www.thefashionlaw.com/home/slaving-in-the-lap-of-luxury-made-in-italy-is-not-what-it-used-to-be. Accessed 2 Feb 2017

Snoy RC (2010) Werkzeugkiste: 23. Fokusgruppen. OrganisationsEntwicklung 29(2):94–98

Spedding E (2016) Met Gala 2016: Emma Watson wears a Calvin Klein dress made from recycled plastic bottles. The Telegraph. http://www.telegraph.co.uk/fashion/events/met-gala-2016-emma-watson-wears-a-calvin-klein-dress-made-from-r/. Accessed 3 Mar 2017

Stella McCartney (2017) POP Eau De Parfum: http://www.stellamccartney.com/Item/Index?suggestion=true&sitecode=STELLAMCCARTNEY_CH&cod10=51121084vr. Accessed 24 Feb 2017

Stewart JS, Oliver EG, Cravens KS, Oishi S (2017) Managing millennials: embracing generational differences. Bus Horiz 60:45–54. doi:10.1016/j.bushor.2016.08.011

There is more to Kering's sustainability report (2016, November 15) The fashion law. http://www.thefashionlaw.com/home/there-is-more-to-kerings-sustainability-report-than-meets-the-eye. Accessed 2 Feb 2017

Timeline: BP oil spill (2010, September 19) BBC News. http://www.bbc.co.uk/news/world-us-canada-10656239. Accessed 2 Feb 2017

Top green companies in the world (2016) Newsweek. http://www.newsweek.com/green-2016/top-green-companies-world-2016. Accessed 2 Feb 2017

Waller DS, Hingorani AG (2014) Luxury organizations and social responsibility: A case study. In: Berghaus B, Müller-Stewens G, Reinecke S (eds) The management of luxury: a practitioner's handbook Kogan. Page Limited, London, pp 423–437

Wiedmann K-P, Hennigs N, Siebels A (2007) Measuring consumers' luxury value proposition: a cross-cultural framework. Acad Mark Sci 7:1–21

Windsor D (2014) Heritage of luxury and responsibility. In: Berghaus B, Müller-Stewens G, Reinecke S (eds) The management of luxury: a practitioner's handbook. Kogan Page Limited, London p, pp 395–407

Winston A (2016) Luxury brands can no longer ignore sustainability. Harvard Business Review. https://hbr.org/2016/02/luxury-brands-can-no-longer-ignore-sustainability

Wittig MC, Sommerrock F, Beil P, Albers M (2014) Rethinking luxury: how to market exclusive products and services in an ever-changing environment. LID Publishing, London

World Commission on Environment and Development (WCED) (1987) Our common future. Oxford, New York

Opal Entrepreneurship: Indigenous Integration of Sustainable Luxury in Coober Pedy

Annette Condello

Abstract In a world of diminishing resources, the opal has become a sign of mineral exclusivity for the consumer luxury market and its value as a luxury object comes from gemstone cognoscenti. According to one Australian Aboriginal legend, rainbow-hued opals are believed by some to stir emotions of loyalty and connection to the earth. Regarding the integral indigenous connection of Australia's national gemstone, rarely has one has looked at the spaces where opal veins were once quarried in remote regions in terms of sustainable luxury. More importantly, the revival of South Australia's opal mining industry in Coober Pedy by female Aboriginal entrepreneur Tottie Bryant in 1946; its development into a multi-million dollar industry into a modern hub in the 1970s; and the spread of the town's construction of subterranean spaces a decade later, enticed immigrants to mine for opals. And when seeking an inexpensive and cool environment, the place enticed immigrants to live underground, providing an unusual form of sustainable luxury in Australia. In 1968, for instance, former Coober Pedy opal entrepreneur John Andrea planned for a unique international underground hotel, the luxurious Desert Cave, but it was not until 1981 when Umberto Coro realised the subterranean spaces' potentiality and created Andrea's dream. Another opal entrepreneur Dennis Ingram designed a golf course with 'scrapes,' which emerged above ground made with opal quarry dust and waste oil. In popular culture too, the town had attracted filmmakers, such as George Miller, to produce his post-apocalyptic epic *Mad Max,* and Wim Wenders, to document his wandering scenes not because of opal scarcity but due to the harsh desert-landscape littered with spoil heaps. Turning to adaptive reuse and indigenous culture in Coober Pedy, this chapter addresses the existing underground passages as the recyclable-integration of a former mining site. In tracking the way in which the community and its rural groundwork served as a site for an innovation in sustainable luxury, the remote underground passages has revealed an unusual Australian lifestyle. Concentrating on the underground spaces, the chapter tracks the manner in which the abandoned sites serve as poignant opal connections within Coober Pedy's integration of remnant spaces and their adaptive

A. Condello (✉)
School of the Built Environment, Curtin University, Perth, Australia
e-mail: a.condello@curtin.edu.au

© Springer Nature Singapore Pte Ltd. 2018
M. A. Gardetti and S. S. Muthu (eds.), *Sustainable Luxury, Entrepreneurship,
and Innovation*, Environmental Footprints and Eco-design of Products
and Processes, https://doi.org/10.1007/978-981-10-6716-7_7

reuse into museums. Opal museums of the future will become magnetic as tourist destinations and their conversion of remnant spaces also into educational facilities foresees the uniqueness of sustainable luxury through its existing empty quarries.

1 Introduction

Today luxury cannot exist without sustainability. 'Sustainability and luxury go hand-in-hand, and will only continue to forge a stronger relationship' (Maisonrouge 2013: 124), that is, with entrepreneurship. In terms of 'luxury as sustainability' by way of constructing a new way of thinking, in considering human interfaces with thoughtful public spaces more critically, adaptive reuse of buildings can be transformed with ease of adaptability to changing needs (Condello and Lehmann 2016). This is the case with Australia's underground spaces where some of them have been converted from mining spaces for opals to sustainable chambers to sell the opals as luxury objects. Opal entrepreneurship and the sustainable luxury industry are interconnected with the Australian Indigenous community.

According to Miguel Angel Gardetti and Ana Laura Torres there is a close connection between the concept of sustainable luxury and entrepreneurship. For these authors, 'sustainable luxury' comprises of 'craftsmanship, preserving the cultural heritage of different nations' (2013: 55). And for 'helping others to express their deepest values' (Gardetti and Torres 2013: 58). 'Sustainable luxury is the return to the essence of luxury with its ancestral meaning, to the thoughtful purchase, to the artisan manufacturing, to the beauty of materials in its broadest sense and to the respect for social and environmental issues' (Gardetti and Torres 2013: 58). This respect for the 'ancestral' involves the indigenous Australian community and how we might associate sustainable luxury with opal entrepreneurship. Gardetti and Torres' definition of 'sustainable luxury' and its essences for different cultural communities are therefore relevant for preserving the cultural heritage of nations and for offering foresight into the connection of the luxury gemstones with their preceding bonding with the ground.

Opals are a naturally-formed luxury. They are bonded with the landscape's lineaments. Over time, they fracture the rock and show colourful seams in situ. Renowned by some for their inspiring colourful seams, opals are themselves luxury objects and their unique qualities are believed to be used to treat sadness, offer protection and impart good luck. Each colour seam is controlled by the size of the silica content and by the opal's refraction surfaces, of individual light waves passing through the transparent cavities. For others, such as the diamond traders of Africa or India, opals conveyed to the gemstone cognoscenti that purchasing an opal to protect themselves was associated with ill luck. Diamond traders had decided to campaign against them by associating opals with a malevolent reputation, which still continues today. White opal, an achromatic colour without a hue, is mined mostly in Coober Pedy and is a sought-after Australian luxury gemstone.

Solid opals have a shelf-life, whereas opal veneers forged with lesser-value stones, often encased with different metals, and tend to fade with time.

First, this chapter looks at the opal's unusual refracting qualities and historical influences. Then, it reveals how Aboriginal opal mythology was considered a natural phenomenon for the purpose of sustenance, indicating the importance of the landscape. The desert and the deserted landscape has informed indigenous entrepreneurship. Following this, it discusses the origins of opal entrepreneurship, the case of Coober Pedy and its underground remnant spaces. Optimistically, these spaces have unleashed the potential to engage further for creating innovative adaptive reuse of diverse indigenous cultures.

2   The Opal's Complicated Facades

In her account of the colour opal as a conceptual substitute icon of wealth, American artist and writer Emily Royson believes the changeable colour gemstone is 'colourless.' Opals acquire a series of complicated façades, which are useful for uncovering the creative facets of entrepreneurship, particularly its cultural innovation and integration of indigenous luxury.

> Opal is a water-jelly mineral that slips through the cracks of stones....
>
> It's slippery and always looks different. It accumulates in the right conditions, and is valued for its purity....
>
> How does the imperfect jelly harden into the myth of meritocracy and upward mobility? (Roysdon 2008: 18–20)

The opal is somewhat capricious. As Royson suggests, 'the rainbow sheen and colour potpourri—consists of impurities in the silica content' due to the 'bent-ray/refracted light-ness of visibility' (Roysdon 2008: 18). After witnessing someone selling a cache of milky opals at a jewellery re-sale shop on 14th Street, New York, for less that she expected, it prompted Royson to rethink about the excess and exploitation of white or 'milky' opals. Pointing towards unsustainability, and struck by its imperfect water-jelly properties, she questions why its association with wealth has become perilous and valued as a rainbow-hued stone to collect and protect. Yet in some parts of Australia, souvenir-tourist shops have cheapened common opals into flaky trinkets where one can buy opal veneers in the form of singlets, doublets or triplets. Nonetheless, Coober Pedy in South Australia is recognised by many as the world's largest producer of precious opals, specifically the 'milk' (or white) opal which was 'traditionally the finest' (Dunstan 1954: 7). Opals are unpredictable aesthetic-looking stones and they most certainly embody indigenous connections, imbued with 'Australian' luxury.

Conventionally, the opal as a valuable luxury object started a craze in eighteenth-century Europe. Towards the end of that century, the opal craze fell out of fashion because its folklore was associated with contagion and scarcity. These 'stones have life, they can suffer illness, old age and death (Baltrusaitis 1989: 83).

Despite the stone's tainted reputation, once again the opal became a fashionable stone to wear as jewellery up until the 1930s and 1940s when Australia was able to offer an abundant supply to the European market. But this was far less the case in the 1960s and 1990s since other gemstones, specifically diamonds and emeralds, quarried from the Americas and Africa proved to be most popular. Consumers decided to invest in purchasing solid high-quality opals, which are judged by their background, vibrancy of the colour and pattern. In Australia black opals are by far the most highly prized luxury objects because of their association with status and rarity, often found in Queensland, New South Wales and Victoria.

IIn Australia rainbow-hued opals are iconic and symbolic. Their veins are judged by their bioluminescence or brilliance and some opals are considered as enriched pictographic stones. The *koriot* opal, for instance, from outback Queensland is recognized as an Aboriginal art opal because their colourful content resembles Indigenous Australian paintings. As pictographic gemstones, they consist of 'accumulations in their depths. Broken, sliced, their brilliant, smooth surfaces reveal perfected ordered veins and pigmentations, signs of some secret writing, bizarre shapes' (Baltrusaitis 1989: 99). Ultimately, the imperfect veins is what makes them unique and unpredictable, unleashing new opal-vein strata in fractionated directions.

More recently, opals have been given a new virtuous reputation and fortune as some people determined they obtained magical properties. Cartier and Tiffany of New York, for example, have created jewellery lines to become a high-market luxury item again. These companies have elevated and re-introduced Australian opals to be on par with other stones, such as emeralds, but they have neglected the cultural significance or the deeper value of their actual physical derivation. In 1993, the opal was proclaimed by politician, and Governor-General at the time, 'Bill Hayden as Australia's national gemstone' (https://www.dpmc.gov.au/government/australian-national-symbols/australian-national-gemstone. Accessed 15.03.2017), promoting its cultural significance.

As an Australian gemstone, however, when exposed to sunlight the opal is an unsustainable natural material because with time the colour eventually fades. This is especially the case when ones sees visible cracks and crazing within the actual opal. It becomes breakable and yet remains exclusive because of its precious mineral content.

Currently, Australia produces ninety percent of the world's supply of precious opal from the sedimentary rocks within the Great Artesian Basin (Milanez et al. 2013: 430; Merdith 2013: 23). The mining community, however, 'has been struggling over the past 20 years' and 'is largely a cottage industry' (Merdith 2013: 23). That is, carried out by 'individuals rather than large mining companies,' which is not 'subject to the same regulations regarding environmental rehabilitation as for other mining or exploration activities in that state (South Australian Opal Mining Act, 1995),' (quoted in Pedler 2010: 37). Other than Lightning Ridge and White Cliffs in New South Wales, Coober Pedy, Mintabie and Andammoka in South Australia are the major producing opal mining areas.

Irrespective of the opal veneers and unpredictable facets and their fading cracks, instead solid opals are considered 'non-renewable resources' (Milanez et al. 2013: 433). In terms of entrepreneurship, which 'refers to risk taking and is considered as a function' (Uzundid et al. 2014), the harsh environment of the opal mining industry opened up the luxury market since there was a demand for the gem overseas, particular in Europe, America and eventually in Asia. This was especially the case in the 1980s with the Japanese demand for them and in the process increasing Australia's tourism industry in South Australia. Rather than market the thin opal veneers into cheap jewellery lines as an alternative option to sell sustainably luxury, it is the natural Australian landscape and its preservation of quarried spaces in Coober Pedy that has made opal entrepreneurship a 'sustainable' luxury.

3 Aboriginal Opal Mythology and Indigenous Entrepreneurship

Before discussing the origins of opal mining in South Australia, its indigenous discovery and entrepreneurship, it is important to reveal how spiritually-significant opals were, and still are, to the Aborigines as revealed in their dreamtime stories. For instance, when the Creator visited earth to bring harmony the foot touched the ground and the opals sparkled into jelly-colours. Supposedly, according to Aboriginal indigenous lore, 'the opal imprisoned the rainbow in the earth' (Weale 1977: 546). For the Wangkumara people from the Coober Pedy Creek region opals brought the gift of fire:

> The tribe sent a pelican away from their camp to find out about the country up north. The pelican felt ill and landed on a hill. The pelican discovered beautiful opal and started chipping it away to bring back to the people. A spark from the chipping caused the nearby dry grass to catch on fire, which spread slowly back to the camp. The fire captured by the people to cook their meat (Connolly 2012).

This particular Aboriginal lore points towards the physical luxury of the opal delivered to the ground as a natural element, for the purpose of sustenance. Similar to Aboriginal opal mythology, ancient Arab cultures believed opal had fallen from the sky and that the play of colour was 'trapped lightning,' as in the case of the Alhambra's artificial/stalactital cave. Another Aboriginal myth tells us about the opal being 'half serpent and half devil, and that brightly coloured fire within the stone was an attempt to lure them into the devil's lair' (http://crystal-cure.com/article-opal-history-properties.html. Accessed 30 December 2016). Aboriginal rainbow-hued opals are thus believed to stir emotions of loyalty and connection, that is, the connection to the ground as the fire of the desert of Australia's arid interior. Through Aboriginal culture, opal mythology's the link to the ground was, and still is, environmentally and spiritually important.

Indigenous Aboriginal entrepreneurship played a part in forming Australian historical trade links. Since the seventeenth century, 'Aboriginal enterprises and entrepreneurial activity have a long tradition… and "are known as some of the world's oldest recorded business undertakings"' (quoted in Brueckner et al. 2014: 1823). There, Aboriginals traded with Indonesia. By the early twentieth century, restrictions were put in place by the South Australian Government that restricted them from their entrepreneurial and enterprising activities. Such activities 'were suppressed by their colonisers who only in recent decades have been trying to revitalise and stimulate Indigenous economic pursuits' (quoted in Brueckner et al. 2014: 1823). This situation included the suppression of South Australia's Indigenous opal entrepreneurship. Yet for some indigenous people, 'digging deep into the ground is considered dangerous, since it would disturb spirits residing under the ground and bring imbalance' (Naessan 2010: 228). Through time, opal mining changed Indigenous perceptions as a way to celebrate its uniqueness as Australia's luxury gemstone.

From the twentieth century onward, indigenous entrepreneurship has been understood and aligned with promoting the foregrounding of social, community-focused aspects. This is important for considering the socio-cultural value of Australia's valuable gemstone to the integration of sustainable luxury of the natural landscape.

Other than admiring their beauty and folklore/mythology and its link to the landscape, Aborigines had little use of opals. In 1925, the new Colorado prospecting syndicate, consisting of gold prospectors Jim Hutchinson and his son were searching for water and instead found pieces of opal lying on the ground. Originally, the place was recognised as the Stuart Range Opal field which later became Coober Pedy, an Aboriginal word '*kupa piti*' which is thought to mean 'white man in a hole.' During the Great Depression, opal prices plummeted and the production almost stalled (http://australian-creation.weebly.com/-rocks.html. Accessed 25/01/2017). Since opals are formed within the ground there was no need to worship opals as luxury objects as in other cultures since Aborigines treat and respect the sacred landscape as it is—in its natural state—to allow for natural luxury to sustain itself.

4   The Origins of Opal Entrepreneurship in Coober Pedy

Internationally recognised as an opal mining and tourist town, Coober Pedy is an arid region and is considered 'one of the weirdest corners in the world' (Hill 1932: 2). As a unique remote area, this sacred land is rich in minerals and deemed as Australia's most valuable prehistoric fossil sites. Located near the Stuart Range, the town of Coober Pedy lies within the Arckaringa Basin and close to the border of the Great Victoria Desert (Naessan 2010: 217). 'Over millions of years, when most of South Australia was covered by an inland sea, it presented the right conditions for opal formation underground' (https://www.samuseum.sa.gov.au/Upload/OPAL_ProgramsFlyerDL_W.pdf. Accessed 25/01/2017).

For Harman

The only reason for Coober Pedy's existence is opals. It is at the centre of the world's largest opal field. Almost a century of opal mining has given the landscape a lunar look for miles around the town. Piles of excavated sandstone dot the surface as far as the eye can see. It is said that there are some two million holes up to 100 feet deep – and mostly three feet across (1999: 26).

Predominantly situated underground, the town is subjected to frequent dust storms and comprises subterranean hotels, houses and opal shops. Tourism is popular in the area too, with its sustainable leisure ground—an aboveground golf course with its scrapes, which arose above ground created with opal quarry dust and waste oil. Golf is played only at night time because of the extreme hot temperatures, existing as a sustainable luxury pastime. In addition, the place has become an iconic cinematic backdrop with its exposure of the town in films such as *Mad Max* and *Priscilla, Queen of the Desert* of the 1980s and 1990s. More importantly, indigenous entrepreneurship of the opal mining industry did not take place in Coober Pedy until the late 1940s. What is problematic, geologically, is the hollowed-out topography, especially the abandoned spaces or ones on the verge of collapse today.

According to Australian literature,

Much of the economic activity in the region (as well as initial settlement of Euro-Australian invaders) is directly related to the geology, namely quite large deposits of opal. The area was only settled by non-Indigenous people after 1915 when opal was uncovered but traditionally the Indigenous population was western Arabana (*Midaliri*), (Naessan 2010: 217).

Some writings suggest that the German immigrant Johannes Menge was the one who discovered opals in South Australia. Menge 'was a promoter of mineral exploration and development in Australia' (Cooper 2011: 194; O'Neil 2011: 1). Within the first two decades of European settlement, the eccentric visionary of mineral exploration 'discovered opal near Angaston in 1839' (O'Neil 2011: 2). He also named many places in the state, such as Carrara Hill. 'It was near this hill that he found 'common' opal (O'Neil 2011: 3).

Following on from Menge, opals were documented as discovered by gold prospector Willie Hutchinson in 1915, as noted earlier, and Coober Pedy's first opal rush was in 1919. At that time, people began to live underground in 'dugouts,' or crude chambers. Coober Pedy's 'opal mining has been carried out since the early 1920s with major booms in activity during the 1960s and 1970s' (Pedler 2010: 37). One could purchase opals in cafes—meaning they were not treated wholly as an exclusive luxury product but as souvenir trinkets—sliced thrice—and made into cheaper jewellery as affordable objects.

As far as opal entrepreneurship is concerned, one Aboriginal woman, Tottie Bryant, predicted success in Coober Pedy in the 1940s. For Aboriginal writer Jessie Lennon who documented Coober Pedy's oral history, recites that on the Mount Clarence Station road Tottie Bryant 'was shepherding her few sheep with her dog when she kicked over a stone and revealed an opal [in 1946]. This discovery at what became the Eight Mile Field started a rush which opened up an extremely lucrative opal field' (2000: 48). Since no water existed in Coober Pedy, Lennon recounted that 'Aboriginal people needed all their knowledge of any available soaks, rock holes, crab holes, water root trees, water bearing plants...to travel to a region such as this in their traditional manner, living off the land' (Lennon 2000: 62). As a survival strategy where the fields attracted a new opal boom, specifically in the Eighth Mile Field, twenty years later opal mining was established on the edge of the town of an Aboriginal Reserve. Deposits of unearthed dirt transformed the material conditions of the natural landscape in the form of its unusual hollowed-out holes. In comparison to other mining towns, such as Lightning Ridge in New South Wales, Coober Pedy's remnant opal mining spaces present us with hewn-out stony surfaces from the tunnelled passages. Consequently, these hollowed-out tunnels later became innovative spaces for tourists in the form of exhibition spaces and the like.

A decade after Tottie Bryant's magnetic discovery, the 'Olympic Australia' opal was found at Coober Pedy, which at the time was regarded as the world's largest valuable gem. This is important as it placed Australia on the luxury-products map. Opal entrepreneurship was therefore significant for the Coober Pedy's mining industry. Tottie Bryant and her tracking team, including husband Charlie Bryant, were clearly innovative for opal mining. Five decades later, her entrepreneurial experience offered potential employment for the Indigenous Aboriginal community as a whole.

To detect the origins of opal veins through its above ground tracking and direction below ground, the entrepreneurship as instigated by Tottie Bryant aided in promoting Australia's luxury gemstone abroad. Where opals were once found, the remainder excavated spaces have garnered new spaces for living-in. these spaces have permitted the potentiality to create an upsurge in the indigenous integration of sustainable luxury, which is culturally and socially important for the local community. That is, from the conversion and extension of existing excavated tunnels into innovative underground spaces.

5   Indigenous Integration of Sustainable Luxury

However, two decades after Bryant's opal discovery, the underground chambers were simply dugouts inhabited by opal miners. At that time, 'the settlement at Coober Pedy housed about 100 miners' families in crudely dug shelters' (Wells 1968, 164). The houses comprise 'of both above-ground and below-ground components and is a response to the great daytime temperatures in the long summer season' (Noble 2007: 133). To locate the opal seams within the underground rock, water had to be recycled because it was scarce.

Today, many of Coober Pedy's citizens live in underground homes. And like any other town in arid conditions the town comprises a petrol station, supermarket, liquor shops, restaurants, religious facilities and underground luxury hotels to cater for the locals and the tourists. 'Hotels are constructed underground in disused opal mines to provide year-round comfort, escaping the searing summer temperatures in these remote communities' (https://www.cooberpedy.sa.gov.au/webdata/resources/files/Gems_of_Coober_Pedy.pdf. Accessed 25/01/2017).

In the late 1940s, according to Lennon, Coober Pedy 'was not a centralised entity; it was scattered over several hectares and had only a few basic facilities: the store, the water tank, and the dugout post office/bank agency. Miners lived where they worked, at the various fields, some of which were located many kilometres distant' (Lennon 2000: 76). In regard to the Umoona Reserve in this field, it was 'exclusively set aside for local Aboriginal people because they had been dispossessed of their traditional lands' (Lennon 2000: 118). Lennon continues:

> The Umoona Reserve (approximately 2025 hectares) was established in 1959, with Pastor Fred Traeger appointed by the State Welfare Department and the Lutheran Church as administrator. His main duty as set out by the Government was to act as opal buyer and shopkeeper for the Aborigines. In 1975 the Aboriginal community adopted the name Umoona, which means Red Mulga tree, common in the area (2000: 118).

In the 1960s the modernisation of opal mining led to a boom. At that time 'a high proportion of European immigrants saw this as a chance to get rich on their own initiative. Population growth forced development and a new hospital, school and another church appeared in the town' (Lennon 2000: 130).

6 Opal Entrepreneurship and Its Underground Remnant Spaces

Since it was Bryant who initially made the rich opal find and revived its industry, two decades later the Umoona site developed into an underground location, which was considered the largest opal showplace in Australia. And in 1976, the Umoona Council bought the site and was converted into the Umoona Opal Mine and Museum (see Figs. 1 and 2).

By the 1980s, Japanese entrepreneur Takeshi Oyama constructed the opal trading corporation Bentine. At its peak, the Bentine Company employed over 200 people and produced a huge profit annually, above one hundred million. He was instrumental in marketing opal to the Asian market and in the development of Japanese tourism in Australia increased. Oyama also had a deep appreciation and great respect for the indigenous culture of Australia but not in the sense of Aboriginal entrepreneurship or what Bryant had initially discovered or the future of sustainable luxury of the town's underground spaces.

Cinematic imagery and settings made Coober Pedy popular as the place as a desert sustainable luxury destination. German film director Werner Herzog chose the town as the backdrop to *Where the Green Ants Dream* (1983), which recounts Indigenous tribal mysteries. 'The Aborigines resist the incursions of a mining company, which wants to churn up one of their dream sites. In court they establish their title to the terrain by producing a *tjuringa* [a sacred totem] they have exhumed

Fig. 1 Coober Pedy (Phil Whitehouse, 20 August 2003). Accessed 3 July 2017, https://commons. wikimedia.org/wiki/File:Coober_Pedy,_Umoona_Opal_Mine_-_panoramio.jpg

—a home-made sceptre, the repository of their ancestral claim' (Conrad 1999: 34) The film opens and closes with 'unAustralian' scenes: a preview of natures' revenge, provoked when the green ants are dislodged from the site of their dreaming' (Conrad 1999: 34). Portraying rather a bleak view, what the film demonstrated was how the town was considered, socially and culturally unsustainable.

When in reality, the town of Coober Pedy's mineshafts clutter approximately the 5000 km^2 mine fields, which vary in depth (http://11geomegangreensmith.weebly. com/sustainability.html Opal mining Coober Pedy, Geography internal 2015).

> The environmental impacts of opal mining on the area are very significant. There are large piles of dirt and dust scattering the landscape and there is a constant danger of running into an uncovered mine. This poses threats to the wildlife living in that area and if no actions are taken, the safety of these animals is at risk (http://11geomegangreensmith.weebly.com/ sustainability.html Opal mining Coober Pedy, Geography internal 2015).

The destruction of flora and fauna in the area has thus led to the prevalence of unsustainable pot-holes together with the impacts of water pollution as well, affecting the natural landscape. With regard to the town's underground remnant spaces, 'about half the population lives underground, with homes carved into the side of the hills and rooms cut out of the sandstone. The underground house were

Fig. 2 Coober Pedy, Umoona Opal Mine (Frans-Banja Mulder, 4 October 2002). Accessed 3 July 2017, https://commons.wikimedia.org/wiki/File:Coober_Pedy,_Umoona_Opal_Mine_-_pan oramio.jpg

first built by returning soldiers after World War I and who helped found the town after the first opal was discovered in 1915' (Harman 1999: 26).

Despite the destruction of the landscape as an unsustainable practice what is a popular attraction for tourists is its numerous mounds of 'discarded rubble from the mines... [and some might someday] find opal missed by the miners.... Opal mining is banned inside Coober Pedy's town limit, but it does go on. People regularly "extend" their homes, digging out new rooms...' (Harman 1999: 26), which suggests the opportunity of discovering more opal veins within the luxury spaces beyond, underground. Extending the dugout to form more luxury spaces, however, might be perceived as innovative but it is ultimately an unsustainable practice.

Coober Pedy's abandoned mine shafts are not hazardous to human safety but also hazardous for the local fauna. Pedler, for instance, has discussed the detrimental impact of fauna entrapment, especially lizards and snakes, in the town's opal prospecting open shafts and uncapped drill holes. 'Pitted with many thousands of uncapped, abandoned shafts; the legacy of exploratory drilling in search of opal trace by hundreds of opal miners throughout the last thirty years. There is no legislative obligation for prospecting shafts to be filled, covered or rehabilitated following exploration or mining activities (South Australian Opal Mining Act, 1995),' (Pedler 2010: 37). Another problem is the unstable flanks of opal prospecting shafts, which are 'naturally filled by debris that falls from the sides or is deposited by wind or water flow from the surface' (Pedler 2010: 41). Dust also presents a massive problem. In using the opal prospecting method in Australia, it does pose a 'significant threat to local reptile fauna' (Pedler 2010: 42) and is likely to repeat patterns in other Australian opal mining areas.

Indigenous people were involved as labourers mining for opals for but not for their own personal use as such. One crucial problem with opals is

> the issue of the relationship between Indigenous rights and the mining industry activities became more central to understanding industry-community relationships, especially as mining expanded in the 1960s and 1970s into central and northern Australia, areas where there were still large populations of Indigenous people. This sparked ongoing debate about not only questions of title but also questions of community development, workforce training, and royalty flows. In many parts of Australia, Indigenous people constitute the largest single community group impacted by major mining projects' (Eklund 2015: 181).

To preserve opal mining quarries into more of a sustainable luxury practice, 'recent accounts by indigenous scholars and leaders such as Marcia Langton and Noel Pearson, for example, have identified mining as the single most important site for local economic development in many rural and remote communities' (quoted in Eklund 2015: 181). This is particularly the situation when referring to sustainable luxury practice in the sense that the sites offer innovation to the Indigenous community. The conversion of remnant spaces into educational facilities, such as the Umoona Mine and Museum, show how quarries have therefore succeeded as a sustainable luxury venture.

One way or another, opal mining in Coober Pedy has 'lost some of its lustre' (Chester 2012). However, approximately 1000 m belowground, 'some of these cool, multi-room residences areas a spacious and luxurious as any in a big city… The landscape is of upmost importance—the Painted Desert and Lake Eyre'(http://www.australiangeographic.com.au/travel/destinations/2012/10/coober-pedy-outpost-in-the-outback. 25/01/2017).

Imitating the landscape's surface and the opal lamination of jewellery lines, as a whole, Coober Pedy offers a case study of Indigenous integration of sustainable luxury, which has been inspirational for the imagination of the renewal (see Fig. 3) and design of new architectural spaces (see Figs. 4 and 5). This is especially the case for displaying fossilised opals as with Glenn Murcutt and Wendy Lewin's current Australian Opal Centre architectural project at Lightning Ridge, New South Wales. Their design appears to integrate the aesthetics of Coober Pedy's opal mining, as well as White Cliffs', underground dugouts into sustainable architecture. This centre will house *fossilised* opals. Ultimately, the building's underground architectural form encapsulates the Coober Pedy's subterranean dugouts.

One drawback is the remoteness of the town of Coober Pedy. And the innovative entrepreneur's future vision, which might impinge upon the delay in a person accomplishing specific goals, because of the distance in travelling to and from the place to other cities. In addition, there is the exposure of people to extreme weather conditions that might delay some in securing future business ventures.

Fig. 3 An underground sustainable luxury house at Coober Pedy to be converted into a private opal museum, 2016 (Courtesy of Tamara Merino)

Fig. 4 The Australian Opal Centre building planned for construction at the Three Mile Opal Field, Lightning Ridge, NSW. Architects Glenn Murcutt + Wendy Lewin. Rendering by Candalepas Associates (Courtesy of Jenni Brammall from the Australian Opal Centre)

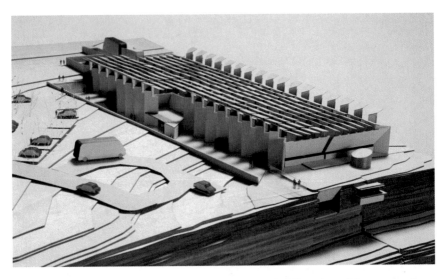

Fig. 5 1:200 scale model of the Australian Opal Centre building planned for construction at Lightning Ridge, NSW. Architects Glenn Murcutt + Wendy Lewin. Model by Scott Choi, Little Models. Photograph by Penelope Clay (Courtesy of Jenni Brammall from the Australian Opal Centre)

7 Opal Quarries as Seamless Spaces of Sustainable Luxury?

The natural properties of the gemstone with its silica spheres and diffracted rainbow hues exceed the crystal structure of diamonds and demonstrates a unique form of national sustainable luxury. Today the Indigenous Jewellery Project, for instance, an initiative that integrates sustainable luxury promotes leadership and craftsmanship. But what does opal mining say about leadership and inspiration?

In respecting indigenous culture and the sacred landscape by integrating and acknowledging the importance and significance of their opal mythology, what is necessary is to map potential areas for future adaptive reuse. That is, to focus on specific areas where opals are found to eliminate wastage.

In terms of the foresight and innovation of opal quarries, presumably we should allow opals to remain in the ground where they are and not treat the gemstones as isolated objects but permit the real opal seams to remain in situ. This is where the real sustainable luxury lies—to integrate the Indigenous historical culture of the place. Inevitably, mapping of the region's existing opal quarries will increase entrepreneurial opportunities in the region as well as for the start-up of alternative initiatives. This will enable the promotion of other sustainable practices associated with opal entrepreneurship to expose a 'seamless' sustainable luxury, naturally, to allow the stony veins to remain where they are.

In valuing Indigenous opal luxury into a more sustainable venture, perhaps the idea of an 'adaptive reuse entrepreneur' would be an apt marketing strategy to promote the futureproofing of the mines. This future outlook might also provide architects and landscape architects or creative environmentalists the opportunity to promote such an intriguing but sustainable underground practice to *innovate* the existing landscapes and to form new concentrated efforts to eliminate man-made wastage.

Above all, as waste, opal veneers are a sustainable option to sell sustainable luxury. And so the natural Australian landscape and its preservation of quarried spaces has enabled opal entrepreneurship to become a sustainable luxury. Natural forms of luxury in the landscape can sustain itself and has created innovative but austere spaces. Internationally, Tottie Bryant therefore promoted Australia's luxury gem industry by creating an upsurge of the cultural and social importance of sustainable luxury through Coober Pedy's underground spaces. Coober Pedy's opal mining booms and wastage have thus transformed into diverse business ventures, which will become innovative for future generations.

References

Australian National Gemstone, Australian Government, Department of the Prime Minister and Cabinet. https://www.dpmc.gov.au/government/australian-national-symbols/australian-national-gemstone. Accessed 15 Mar 2017

Baltrusaitis J (1989) Aberrations: an essay on the legend of forms. MIT Press, Cambridge, Mass

Better World Arts. Cross Cultural Projects. http://www.betterworldarts.com.au/about. Accessed 15 Mar 2017

Bramall J (2015) Australian Opal Centre: a glittering palace for the queen of gems. GAA J 25(9): 304–315

Brueckner M, Spencer R, Wise G, Marika B (2014) Indigenous entrepreneurship: closing the gap on local terms. J Australian Indigenous Issues 17(2):1820–1829

Chester Q (2012) Coober Pedy: outpost in the Australian outback. OCT 29, 2012, Australian Geographic 111, Nov–Dec 2012. http://www.australiangeographic.com.au/travel/destinations/2012/10/coober-pedy-outpost-in-the-outback

Condello A, Lehmann S (eds) (2016) Sustainable Lina: Lina Bo Bardi's Adaptive Reuse Projects. Springer, Switzerland

Conrad P (1999) The second discovery of Australia. LINQ 26(2):28–39

Connolly MJ (2012) Quoted in Chapter 5 'Old, flat and red - Australia's distinctive landscape,' in Shaping a Nation: a geology of Australia, Blewett R (ed.), ANU Press, p. 265

Cooper BJ (2011) Geologists and the Burra Copper boom, South Australia, 1845-1851. Hist Res Mineral Res 13:193–200

Dunstan P (1954, MAy) Dugout life. The Sydney Morning Herald Newspaper, Monday 17: 7

Eklund E (2015) Mining in Australia: an historical survey of industry-community relationships. Ext Ind Soc 2:177–188

Gardetti MG, Torres AL (2013, December) Entrepreneurship, innovation and luxury. J Corporate Citizenship 55–75

Harman A (1999, Mar) Coober Pedy policing: things have quieted down in Australia's 'wild west.' Law and Order 47(3): 26–29

Hill E (1932, June) 'It's all luck on the Opal fields,' The Mail Newspaper, Adelaide, Saturday 25: 2

Layman L 'Mining,' Murdoch University. http://www.womenaustralia.info/leaders/biogs/WLE0382b.htm. Accessed 30 Jan 2017

Lennon J (2000) I'm the One that Know this Country. Aboriginal Studies Press, Canberra, Australian Capital Territory

Maisonrouge KPS (2013) The Luxury Alchemist. Assouline, USA

Merdith AS et al (2013) Towards a predictive model for opal exploration using a spatio-temporal data mining approach. Australia J Earth Sci 60:217–229

Milanez B (2013) Innovation for sustainable development in artisanal mining: advances in a cluster of opal mining in Brazil. Resources Policy 38:427–434

Naessan P (2010) The etymology of Coober Pedy, South Australia. Aboriginal Hist 34:217–233

Noble AG (2007) Traditional Buildings: a global survey of structural forms and cultural functions. IB Tauris, London, New York

O'Neil B (2011) Johannes Menge: more than 'father of South Australia's Minerology,' Professional Historians Association: 1–3

Pedler, RD (2010, April) The impacts of abandoned mining shafts: fauna entrapment in opal prospecting shafts at Cooer Pedy. South Australia, Ecolog Manage Rest 11(1): 36–42

Petter N (2010) 'The etymology of Coober Pedy. South Australia', Aboriginal History 34:217–233

Royson, E (2008) Colors/Opal, Cabinet Magazine, 29, Spring: 18–19

The Indigenous Jewellery Project. http://aboriginaljewellery.com.au/

Uzunidis D et al (2014) The entrepreneur's 'resource potential' and the organic square of entrepreneurship: definition and application of the French case. J Inn Entrepreneurship 3(1): 1–17

Weale R (1977, June) Opal: the rainbow gem, in New Scientist, No.1054, 2

Sustainable Luxury Tourism, Indigenous Communities and Governance

Anne Poelina and Johan Nordensvard

Abstract Sustainable luxury cannot only be understood as a vehicle for more respect for the environment and social development, but also as a synonym of culture, art and innovation of different nationalities and the maintenance of the legacy of local craftsmanship. The overall aim of this chapter is to explore the important intersection between traditional Aboriginal cultural and environmental management, knowledge and heritage, with the interest of sustainable luxury tourism in remote wilderness communities in Australia. Socially Sustainable luxury tourism could encompass important element of empowering and life-sustaining activities for remote Indigenous groups on a global scale based if informed by Indigenous cultural governance to facilitate sustainable tourism. We argue that such a development could bridge the divide between culture and nature explaining how and why management and protection of landscapes and eco-systems are integral to human heritage, culture and a new wave of sustainable luxury tourism. The Mardoowarra, the Fitzroy River and its life ways, in the vast Kimberley, northern Western Australia, is highlighted to exemplify both our meaning and concern.

Keywords Sustainable luxury tourism · Indigenous communities
Indigenous governance

A. Poelina (✉)
Nulungu Research Institute, The University of Notre Dame Australia,
88 Guy St., Broome, WA 6725, Australia
e-mail: majala@wn.com.au

J. Nordensvard
Sociology, Social Policy and Criminology, University of Southampton,
Highfield, Southampton SO17 1BJ, UK
e-mail: j.o.nordensvard@soton.ac.uk

1   Introduction

The global population of Indigenous[1] people still suffers today from discrimination, marginalisation, extreme poverty and conflict. Indigenous Australian people have and are still facing overall social, economic, juridical and political disadvantages in Australian society (Aboriginal and Torres Strait Islander Affairs 2003; Ivory 2003; Fuller et al. 2005; Pink and Allbon 2008; SCRGSP 2016). This illustrates higher unemployment vis-a-vis non-Indigenous groups, and that average income and rates of business ownership are significantly lower than among non-Indigenous groups (Aboriginal and Torres Strait Islander Affairs 2003; Fuller et al. 2005; Pink and Allbon 2008; SCRGSP 2016). There are major challenges to remove barriers of discrimination, and to access and increase the quality of social services available to Indigenous communities.

A large part of the Indigenous population lives in regional and remote areas with fewer prospects of employment, which has led to an interest from Australian government agencies to regard tourism as a viable route to promote Indigenous entrepreneurship and employment (Commonwealth of Australia 2010; Hinch and Butler 2007; SCRGSP 2016). A major priority has been to increase tourism investment in regional and remote areas through a diverse array of Federal and State government programmes (Buultjens et al. 2005; Ivory 2003; Whitford and Ruhanen 2009).

There are many obstacles such as lack of land tenure, difficulties in raising finance, the design of tourist itineraries, and a lack of market profile and market skills that have undermined and are still undermining the prospects of Indigenous tourism (Altman 1993; Altman and Finlayson 2003; Dyer et al. 2003; Zeppel 2001). Schmiechen and Boyle arge that Indigenous tourism "remains an extremely fragile and tenuous sector of the tourism industry" (2007: 60). Higgins-Desbiolles et al. (2014) have highlighted the importance of including and incorporating an understanding of Indigenous culture and Indigenous community aspirations.

Socially Sustainable luxury tourism could become an important part of empowering and life-sustaining activities for remote Indigenous groups and unique habitats such as the Mardoowarra, Kimberley, and its unique environmental assets. Sustainable luxury can not only be understood as a vehicle for more respect for the environment and social development, but also as a synonym of culture, art and innovation of different nationalities and maintaining the legacy of local craftsmanship (Gardetti 2011), which becomes even more important for Australian Indigenous culture and communities.

Our argument in this chapter that the ethos and practice of sustainable luxury tourism could be a vehicle to incorporate an ethic of care for visited places (Mair and Laing 2013; Miller et al. 2010; Walker and Moscardo 2014) while providing

---

[1]Throughout this chapter the words Aboriginal and Indigenous are capitalised as they are proper nouns and are used interchangeably to describe First Nations peoples of the Australian mainland and Torres Strait Islands.

Indigenous employment. This could be achieved through a focus on small and community-based sustainable luxury tourism aimed at overseas visitors. This might be essential in developing Indigenous tourism to combat some of the fundamental problems such as remoteness (access to resources/travel costs of tourism). Mass tourism tends not to benefit local communities and there is a lack of interest from the nation's tourist markets.

We will redefine what we mean as luxury in sustainable luxury, focusing on the uniqueness of access to a spiritual and environmental experience created through the Indigenous understanding of the environment. The Kimberley region in the North West of Western Australia, our primary focus, is considered to be one of the last great wilderness areas of the world, and it was listed as an area of National Heritage in 2011. We argue that such development can bridge the divide between culture and nature and explain how and why management and protection of landscapes and eco-systems are integral elements of human heritage, culture and a new wave of sustainable luxury tourism. The value of Indigenous governance and the rights to their land is central to building a case for sustainable luxury, and to promoting multi-layered ethical care of the places people visit and of becoming part of both the protection of the land and their cultures.

The chapter will be divided into four main parts. The first part will provide the background around the concepts and the chosen case study. The second will briefly discuss the issue of land rights for Indigenous people. The third part will discuss the importance of Indigenous governance. The fourth will aim to analyse the potential for sustainable luxury tourism for Indigenous communities and then conclusions will be drawn.

2 Background

2.1 Mardoowarra (Fitzroy River)

The Mardoowarra (Fitzroy River) is the mightiest river in the Kimberley with a length of 733 km, and when in flood during the wet season, it carries so much water that it is second only to the Amazon River for volume (Watson et al. 2011). Mardoowarra, or the Fitzroy River, Valley Tract, is located along the contact of the Kimberley Region and Great Sandy Desert. "The Fitzroy River empties into King Sound to form the largest tide-dominated delta in the World. The total valley from Fitzroy Crossing to King Sound is globally unique in its biotic and abiotic diversity, remoteness, and arid setting" (Semeniuk and Brocx 2011: 151–160). The river originates in the King Leopold Range and comprises a vast catchment area of almost 100,000 km². This encompasses more than 20% of the Kimberley region. The river is up to 1 km wide in parts, and is flanked by a 300 km long floodplain spanning up to 15 km wide, before finally emptying into King Sound south of Derby. The Mardoowarra is one of the most mega-diverse regions in the world from a biogeographic perspective (Pepper et al. 2014).

Vogwill highlights how the pristine river cuts through a variety of ancient terrain, including sandstone plateaux, and 350 million-year-old Devonian limestone of former barrier reefs that has eroded into deep and dramatic gorges. Vogwill highlights further significant fauna in and around the river which includes 18 endemic fish species found nowhere else in the world such as the endangered freshwater Sawfish and freshwater crocodiles, sharks, rays, turtles, mussels, waterbirds, falcons, bats and quolls (Vogwill 2015: 6–11). The fluvial vegetation, such as freshwater mangroves, also provides rich sources of food and traditional medicines (Watson et al. 2011).

In 2011, the Australian Government announced the National Heritage Listing of 19 million ha of the West Kimberley including the entire Mardoowarra for its outstanding cultural significance to the nation. In recognition of Mardoowarra's outstanding cultural value and connectivity of the river system, on 31 August 2011 the Federal Minister for Environment and Heritage included the river as part of the West Kimberley National Heritage area stating "the Fitzroy River and a number of its tributaries, together with their floodplains and the jila sites (waterholes) demonstrate four distinct but complementary expressions of the Rainbow Serpent (Yoongoorrookoo) tradition associated with Indigenous interpretations of the different ways in which water flows within the catchment and are of outstanding heritage value to the nation under criterion (d) for their exceptional ability to convey the connectivity of the Rainbow Serpent tradition within a single freshwater hydrological system" (Commonwealth of Australia 2011).

The exceptional natural and cultural value of the river not only added a qualitative dimension to the Kimberley and Western Australia, but also to the nation. The entire Fitzroy River catchment was added to the National Heritage Listing in 2011 by the Australian Government. The Fitzroy River is also listed as an Aboriginal Heritage Site under the Western Australian Aboriginal Heritage Act 1972.

2.2   Social Innovation

Innovation systems are highly complex and need to reconcile socio-economic development, environmental sustainability and technological and innovation capabilities, in both developed and developing countries (Stamm et al. 2009; Urban et al. 2012). Social innovation has become a separate stream of literature that focuses on the social and political organization of society.

Ockwell and Mallett regard 'social innovation' as "recognizing new ways of organizing or doing things through social dimensions" (2013: 118). Andersen et al. define social innovation as "the ability to organise bottom linked collective action/empowerment (including efficient political representation)", and state that this "is a condition for reaching sustainable democratic and social development" (2009: 283). They argue further that "[S]ocial inclusion and integration are impossible without both social conflict and democratic dialogue" (2009: 283). We argue that sustainable luxury tourism that is socially just needs to take into account redistributive, procedural justice and in particular inclusive participation and rights and responsibilities under guardianship and custodial considerations, along with key principles of Indigenous ownership of both governance process and the benefits of development.

We argue that a transfer towards small-scale sustainable luxury tourism and Indigenous governance promoting David corporations and overseas guests could be a disruptive innovation (Christensen 1997). Disruptive innovation could help create new markets and potentially disrupt the mainstream tourism market by implementing different value sets (Christensen 1997; Christensen and Raynor 2003; Christensen and Overdorf 2004). Sustainable luxury tourism based around Indigenous environmental governance could add an important and educational element to visitors engaging these types of experiences in remote communities.

2.3 Sustainability

Sustainable development and environmental management discourses try to bridge tensions between humans and the habitats that humans live in. Kearin, Collins and Tregidga argue that sustainability is "a systems concept that has at its heart ecological sustainability and the longevity of biophysical systems that support human life." (2010: 519) Sustainability was popularized by the Brundtland Report in 1987 which became synonymous with "development that meets the needs of the present without compromising the ability of future generations to meet their own needs" (World Commission on Environment and Development 1987: 43).

There is an interest in integrating cultural, social, economic and environmental dimensions of development. These dimensions are regarded as the core pillars of sustainable development, and sustainability has become one of the most important environmental concepts in development. Still, the social pillar was often seen as a lower priority than both the environmental and economic dimensions. There have been ongoing attempts to link social sustainability to the other dimensions of sustainable development and wider policy issues (Littig and Grießler 2005; Davidson 2009; Dillard et al. 2009; Casula et al. 2012; Dempsey et al. 2011).

Marcuse argues that sustainability should not be thought of as conceptualising the current global status quo with all of its inequalities (1998). There has been an attempt to link social sustainability to the concept of environmental justice. Agyeman and Evans argue that "just sustainability" needs a clear linkage between sustainable development and environmental justice to prevent the social pillar from becoming one-sided (2004). Harvey points out that environmental injustices need to be a first priority on the sustainability agenda (Harvey 1996: 385 ff).

Schlosberg and Carruthers stress that Indigenous demands for environmental justice are not just about distributional equity but also the functioning of Indigenous communities, which highlights traditions, practices, and protecting the essential relationship between Indigenous people and their ancestral lands and waters (2010). The link to the capability/functioning approach of Sen and Nussbaum has also meant an expansion of the original scope on individuals towards the functioning and capabilities of Indigenous communities and their environment (Schlosberg and Carruthers 2010).

Schlosberg argues further of the need to add a capability dimension to the environment in environmental justice. This would "enrich conceptions of environmental and climate justice by bringing recognition to the functioning of these systems, in addition to those who live within and depend on them" (Schlosberg 2013: 44). Tourism based on Indigenous governance in Australia will go further than just seeing places, but understanding the underlying relationships between humans, culture and the environment in one of the oldest continuously existing cultures.

2.4   Why Sustainable Luxury Might Play an Important Role

Luxury is seen by De Barnier as having seven culturally specific common characteristic elements, which are exceptional quality; hedonism (beauty and pleasure); price (expensive); rarity (which is not scarcity); selective distribution and associated personalised services; exclusive character (prestige, privilege) and creativity (art and avant-garde) (De Barnier et al. 2012). Luxury has been defined by Klaus Heine (2011) as something desirable which goes beyond being merely a necessity. Kapferer and Basten depict luxury to signify prosperity, power and social status (2010) and luxury is constant in change and reflects the social norms and aspirations of different times and societies (Berry 1994).

The term 'luxury' must not be seen as opposing ideals or ethics but could actually be turned into ethical production and consumption. "Luxury is also associated with high quality, know-how, slow time, the preservation of hand-made traditions, transmission from generation to generation of timeless products: these associations will be in agreement with sustainability" (Kapferer and Michaut 2015: 5). A luxury strategy often involves locally-produced products that respect sources of raw materials (Kapferer 2010). Problems have occurred when prestigious brands have abandoned the luxury route for second and third tier products produced often under similar but non-luxury conditions. As the world faces overconsumption, there is an imperative that we consume less of lesser quality and more of higher quality, in terms of shared and valued experiences.

There has been discussion of how some emerging disruptive and innovative companies within the luxury industry embrace values in a more fundamental way (Bendell 2012). Often defined as Davids against Goliaths, these companies are based on a pronounced value approach that aims to support social and environmental changes (Hockerts and Wüstenhagen 2009). Traditional luxury companies have encountered problems in implementing sustainability in their business and in their products (Bendell and Kleanthous 2007) Jen Morgan expressed this when she stated that "[t]o leapfrog ahead, we need pioneering and brave people, communities and organizations who are willing and able to challenge that status quo and to experiment for change" (Morgan cited in Gardetti 2014: 32). Sustainable luxury is an important endeavour to integrate the knowledge and craftsmanship of Indigenous people to ensure that Indigenous communities and future environmental capabilities are supported.

2.5 Indigenous Tourism in Australia

Indigenous groups in Australia face higher unemployment vis-a-vis non-Indigenous groups and the average income and rate of business ownership is significantly lower than among non-Indigenous groups, which explains why tourism is seen as a development tool and a way to create "much needed opportunities for employment, social stability and preservation of culture and traditions" (Commonwealth of Australia 2003: 41). Tourism has become a viable option for Indigenous people to establish themselves in the economy (Fuller et al. 2001).

Cultural tourism has been regarded as important as it would involve the land and Indigenous cultural assets, but Altman and Finlayson (1992) highlight the importance of balancing cultural integrity with concepts of commercialisation. Therefore it is important to create tourism centred around Indigenous governance to prevent destructive intrusions, invasions of privacy and trespassing.

There are many challenges and opportunities facing Indigenous sustainable luxury tourism and the remainder of the chapter discusses some of the more significant issues and topics. Fletcher et al. (2016), for instance claims that there are three major challenges that could undermine Indigenous tourism endeavours, such as control or security of tenure and recognition of legal rights to ancestral or traditional lands on one side, and the management of the land on the other; the overall governance situation, and finally the policy context. Concentrating on the Mardoowarra as an example, our discussion is furthered developed.

3 Control, Tenure and Legal Rights Vis-a-Vis Management Responsibilities

Coria and Calfucura (2012), Bunten (2010), Weaver (2009) and Colton and Whitney-Squire (2010) have argued "that control or security of tenure and recognition of legal rights to ancestral or traditional lands and waters is a key issue influencing the success of Indigenous tourism enterprises" (Fletcher et al. 2016: 1106). Coria and Calfucra (2012) argue that if Indigenous communities cannot assert control over land and resources, investment in tourism is prevented, and has the potential to limit the possible social and economic benefits of such projects. Langton et al. (2005) argue that such "guaranteed land security and the ability of Indigenous and local peoples to exercise their own governance structures is central" to maintaining traditional knowledge systems upon which Indigenous livelihoods depend (24).

The recognition of Native Title in 1993 following the momentous Australian High Court Mabo decision in 1992, was a significant advance in the position of Indigenous people, as their rights to land and waters was interpreted by Australian law on the basis that it recognised that prior to colonisation Indigenous Australians had their own laws and customs. It is important for Indigenous people to build upon these rights as recognised within the *Native Title Act* to ensure all Indigenous peoples can benefit from the commercial use of traditional lands and waters. Furthermore, traditional owners have an inherent right to make decisions about customary cultural and native title rights and responsibilities on and for country. Tran (2015) champions this right in accordance with Article 19 of the United Nations Declaration on the Rights of Indigenous Peoples (UNDRIP 2007), which states that any policies and legislation developed need to ensure that Indigenous rights and responsibilities towards guardianship and custodianship for their lands and living waters are paramount. Article 26 of the United Nations Declaration on the Rights of Indigenous Peoples states that:

> Indigenous peoples, have the right to own, use, develop and control the lands, territories and resources that we possess by reason of traditional ownership or other traditional occupation or use, as well as those which we have otherwise acquired.
>
> States shall give legal recognition and protection to these lands, territories and resources. Such recognition shall be conducted with due respect to the customs, traditions and land tenure systems of the Indigenous peoples concerned.

Tran (2015) asserts that in accordance with Article 32 of the United Nations Declaration on the Rights of Indigenous Peoples (2007), that:

> States shall consult through our representative institutions in order to obtain our free and informed consent prior to the approval of any project affecting our lands or territories and other resources, particularly in connection with the development, utilisation or exploitation of mineral, water or other resources.
>
> States shall provide effective mechanisms for just and fair redress for any such activities, and appropriate measures shall be taken to mitigate adverse environmental, economic, social, cultural or spiritual impact.

All negotiations and decision-making agreements must be engaged through a free, prior and informed consent. It is therefore paramount that Aboriginal people must have a central role in the development, implementation and evaluation of policy and legislative or administrative measures that impact on their lives.

A right to sustainable life and sustainable development must be grounded in an Aboriginal context, and it is important to increase an understanding around complex relationships of sites, or land, or country, which cannot be separated from the people, or custodians, who live there and care for them. These principles of justice can lead to an improvement of living standards and allocation of resources to rural and remote communities in a more entrepreneurial, economically sound and environmentally just way.

In 2011, the Australian Government announced the National Heritage Listing of West Kimberley, including the entire Mardoowarra. The Australian Government has recognised that Indigenous people have "long engaged in productive activities in and around wild rivers, with a deep knowledge base, awareness and attachment to the life of the river" (Wells 2015: 7). Mardoowarra (Fitzroy River) is one of Australia's few remaining wild rivers, "relatively unaltered by modern human development, and exist[ing] in [its] natural condition—to flow freely without dams or other barriers" according to Wells (2015: 7) On the 2nd and 3rd of November 2016, guardians and custodians of the Fitzroy River Catchment met in Fitzroy Crossing to champion the Fitzroy River Declaration (2016). The Declaration is the mechanism for Indigenous governance to develop a Management Plan to respond to the growing concerns of the extensive development proposals facing the Fitzroy River. The unique cultural and environmental values of the Fitzroy River and its catchment are of national and international significance. The Fitzroy River Declaration (2016) sets a national standard for Native Title as well as enshrines the UN Declaration on the Rights of Indigenous Peoples for self-determining responsibilities as guardians of the Fitzroy River as being fundamental to the management of this globally unique river system.

The Declarations send a strong message to the Federal Government to endorse the Environmental Protection and Biodiversity Conservation (EPBC) Act (1999) Draft Referral Guidelines for the West Kimberley National Heritage Places (2012) as the guiding principles for development within the Fitzroy Catchment. The Declaration explicitly encompasses the diversity of Aboriginal peoples, groups and communities who live along the Mardoowarra (Fitzroy River) and continue to maintain a special relationship with this sacred river. The Fitzroy River is also listed as an Aboriginal Heritage Site under the Western Australian Aboriginal Heritage Act 1972. The guardians and custodians believe the Fitzroy River is a living entity and an ancestral being and has a right to life. It must be protected for current and future generations, and managed jointly by the Traditional Owners of the river. What binds Kimberley's Indigenous peoples is the stories and wisdom and the collective and continuing responsibility the guardians and custodians have to maintain custody and guardianship of the Mardoowarra/Fitzroy River as an asset in Common, registered and endorsed as National Heritage (Poelina and McDuffie 2016b). The Indigenous governance approach is regulated under Aboriginal law and an Indigenous management regime that expects that both cultural and legal responsibilities on the country are implemented, evaluated and that the learning is shared.

At the same time, the Mardoowarra is under threat from direct and indirect development to transform sections of the river into intensive farming areas, with large scale land clearing and water extraction, and there is the ever-growing threat of coal mining and gas fracking in the Mardoowarra catchment area (Wilderness Society 2017). Using Indigenous tourism products in Mardoowarra will provide not only an ethical and commercial support of Indigenous native title rights, but is also supporting Indigenous communities' management of the environment in a sustainable way.

4 Indigenous Governance

There has to be a discussion of how conceptualising and implementing an Indigenous governance within sustainable tourism could support positive returns and minimise negative effects through long-term strategic planning, stakeholder collaboration and community empowerment (Whitford and Ruhanen 2010). Bennett et al. (2012), Colton and Whitney-Squire (2010) and Bunten (2010) emphasise that governance processes provide a cultural authority with the means of ensuring legal recognition, accountability, inclusiveness, participation and conflict resolution with the presence of formalised bodies supporting economic development, access to and decisions on the use of land, living waters and natural resources. There is a need for governments to take a new approach to researching, planning and development which impacts on the right to life for the Mardoowarra. This relationship must be a partnership based on principles that recognise the continuing cultural and economic rights of Aboriginal people, a commitment to democratic process, and to improving leadership, governance and entrepreneurial capacity.

There are attempts to link Indigenous entrepreneurs as described in Freire's discussion of praxis: transforming situations through action and critical reflection (Freire 2005). In his dialogic action theory, Freire distinguishes between dialogical actions which promote understanding, cultural creation, and liberation; and non-dialogic actions, which deny dialogue, distort communication, and reproduce oppressive power structures. The 'strength-based' approach rejects narratives that promulgate inferiority. This approach has a greater focus on innate ability, the advantages of Indigenous culture (rather than framing it as disadvantageous), dialogue, and 'hopes and aspirations' for 'how we want to be' (2005: 168). Research that employs this strengths-based approach is consistent with Smith's (1999) call for Indigenous communities to 'reframe' how Indigenous nations are connected physically, culturally, spiritually, entrepreneurially and economically to re-build their cultural governance relationships in their protection and sharing of cultural and environmental knowledge, practice and assets, to ultimately promote sustainable life and sustainable development across the generations and across the world.

It is vital to see Indigenous tourism as going further than sustainable development and seeing the need to create sustainable life. Poelina discusses how one needs to link the rights of human beings with the right to life for nature (Poelina and McDuffie 2016a). One could link such an approach to a more extensive understanding of environmental justice which also concerns the functioning of Indigenous communities, which highlights traditions and practices, and protects the essential relationship between Indigenous people and their ancestral lands and waters (Schlosberg and Carruthers 2010).

By utilising the principles of Indigenous Nation Building, the evidence emerging in Australia suggests that culture (the beliefs, practices, and ethics of law and custom) is the mechanism that Indigenous peoples use to participate in the world around them, or to 'be with' and 'act as' guardians for their tribal lands and living waters. Regional and collaborative governance must be determined by the people most affected. Indigenous peoples are key stakeholders in such partnerships and come to share their lived and rich experiences from their deep inter-generational relationship with nature. Sustainability is very much a part of Indigenous governance. According to the Northern Australian Indigenous Land and Sea Management Alliance (NAILSMA 2009):

> Indigenous Peoples have rights, responsibilities and obligations in accordance with their customary laws and traditions, protocol and customs to protect, conserve and maintain ecosystems in their natural state so as to ensure the sustainability of the whole system.

If Indigenous nations are to rebuild enduring governance and wellbeing; on the social, cultural, human and environmental assets these types of entrepreneurial, innovative ways of doing business provides their citizens and their nations. It is the process of dialogue to action through collaborations with others to build trust, sustainable tourism, environmental protection and justice.

5 Overall Policy Context for Sustainable Luxury Tourism

Much of the dominant discourse around Indigenous tourism in Australia has been based around marketing, product development and economic benefits as to why Indigenous tourism development should be supported (Whitford and Ruhanen 2010). Few have paid attention to Quality of Life of Indigenous communities or to their role as environmental stewards (Whitford and Ruhanen 2010). If we look at the different factors presented earlier such as both tourism as a way to combat social exclusion, strengthen the capabilities of Indigenous remote communities and protect the environment and eco-systems, it becomes apparent that a classical understanding of tourism might fall short in reaching these goals. Sustainable luxury tourism opens up new possibilities in reaching these goals.

5.1 High Costs

One of the many challenges that Indigenous tourism faces is often the discrepancy between the perception of a high demand for Indigenous tourism (Department of Industry, Tourism and Resources 2005; Tourism Research Australia 2010) and actual national demand. The potential market for Indigenous tourism is predominantly international visitors, which means that visitors will go to some lengths both in terms of cost and transport to reach remote Indigenous communities. This remoteness also leads to higher costs such as remoteness from suppliers and markets (Young 1988). There are other issues such as access to start up and developmental capital (Finlayson and Madden 1995; Fiszbein 1997), and government schemes have been difficult for Indigenous people to access (Buultjens et al. 2002).

This becomes even more salient if many visitors were international rather than national. "Conventional wisdom is that Indigenous tourism is much more popular among international tourists, especially from Northern European countries, than among domestic tourists. Moreover, Indigenous ecotourism enterprises are to some degree protected from competition due to their relative remoteness" (Coria and Calfucura 2012: 51). Ryan and Huyton have found in a study of visitors to Katherine that only about a third of tourists indicated a high level of interest in Aboriginal tourism products, whereas those who are younger, often female, better-educated and from North America or Northern Europe indicated high interest (2002). This argument highlights the fact that sustainable luxury tourism might not just be desirable, but also necessary.

To create Indigenous tourism that is both sustainable and lucrative according to high ethical, social and environmental standards is challenging as it goes beyond our standard understandings and experiences of tourism, implying that the ethos of sustainable tourism could be an option to inculcate an ethic of care for visited places (Mair and Laing 2013; Miller et al. 2010; Walker and Moscardo 2014).

5.2 Redefining Luxury

Luxury does not have to mean a standardised understanding of comfort and service, but unique and bespoke itineraries that are constructed with environmental and spiritual knowledge of Indigenous communities and entrepreneurs. If people travel to eat luxury degustation [A1] menus in remote areas such as Fäviken in Sweden or Awasi in Chile, there are good arguments for understanding Indigenous tourist experiences as degustation experiences in remote and unparalleled wilderness. Just as in Haute Cuisine restaurants, much effort is put into the craftsmanship and use of resources. Indigenous governance and management are similar in managing the environmental resources in an ethical and sustainable way. Indigenous management and governance should therefore be seen as central elements of any tourism products in remote areas.

In many ways Indigenous governance and management of the Mardoowarra could work if it emphasised sustainable luxury tourism as part of a society of experience (Erlebnisgesellschaft). People do not see their lives as part of a struggle to survive, to follow duties and principles from a divine source, but as a search for variety, interesting experiences and self-fulfilment (Selbstverwirklichung). Consumption and communications are the main lines of this new search for identity and self-realisation (Ludes 1997: 89). Girón sees in luxury a human striving for beauty, refinement, innovation, purity, the well-made and the aspiration for perfection and what is the best experience (2012).

It is therefore important to reconnect to artisanal luxury craftsmanship and a move from quantity of consumption towards higher quality in a time that mass-consumption is becoming highly unsustainable. There have therefore been attempts to rethink both the nature and goal of luxury, such as Bendell and Kleanthous (2007) who argue that luxury products and services are intertwined by both consumers' aspirations and could embody values such as environmental and social issues. Gardetti argues that to "achieve a profound social change, the role of personal values is very important: idealistic values regarding environmental and social goals can be translated into value economic assets" (2014: 26) The values expressed in the two previous sections on Indigenous Native Title and Indigenous governance of the Mardoowarra, express values that the tourist could align to and also use to find their own identity and self-realisation.

5.3 Scaling Down

According to Mowforth and Munt (1998, 1277), "if sustainable tourism policies and measures are not established early on to manage the possible negative effects of tourism, initial tourism development can become a political and marketing gimmick that opens the door to unwelcome mass tourism". Two cases from China have shown that for Indigenous communities, they need to be involved if the tourism is to have any substantive impact on reducing poverty. An interesting example is by comparing Indigenous tourism in Yunnan and Guizhou, where tourism has been promoted on a large scale in already relatively well-off areas in the former, whereas the latter has focused on low-key tourism in relatively poor areas. "Overall, in Guizhou, the distribution and structure of the tourism industry contributed directly to reducing rural poverty in the province to a greater extent contributed directly to reducing rural poverty in the province to a greater extent than it did to economic growth" where "Yunnan's extensive tourism industry, by contrast, promoted the province's rapid growth, while contributing surprisingly little towards eliminating Yunnan's poverty" (Donaldson 2007: 36).

The structure of Guizhou's tourism encourages the participation of the poor, while in Yunnan poor people are often excluded. The focus on upscaling has ensured higher economic growth in already non-poor areas, whereas smaller groups of tourists have contributed less to economic growth but more towards poverty reduction in poor rural communities that directly participate in tourism. This case shows that focusing on smaller groups of tourists where the benefits go directly to remote communities is more desirable than large tourist developments.

5.4 Indigenous Entrepreneurs

Tourism is seen as income generation for Indigenous communities that demand relatively low levels of government intervention and support. Often this could Indigenous participation in the hospitality and retail services area, in cultural, safari, wilderness and bush tucker tours or in making and selling arts and crafts. Moreover, Indigenous could be employed through their organisations to facilitate tourism (Fuller et al. 2005: 892–893). This shies away from the fact that the Indigenous tourism undertaken that is nation-building, increases indigenous capacity and protects the environment will not be a mass-produced, mainstream endeavour.

There is a good argument that we will need to redefine how we understand Indigenous tourism as a high-end product with high costs for both entrepreneurs and visiting tourists. It becomes important to discuss how Indigenous innovation and entrepreneurship could be promoted to "break the rules" and to "promote disruptive solutions" to both environmental and social issues. To understand social change one needs to start with understanding the role of personal values. One could assume that idealistic values concerning both the environment and social goals could potentially be translated into value economic assets (Dixon and Clifford 2007).

In the best cases, Bennett et al. (2012), Bunten (2010), Colton and Whitney-Squire (2010), Fuller et al. (2005) and Turner et al. (2012) highlight the fact that tourism can build capacity, foster the integration of economic, social, cultural and environmental objectives and support Indigenous community development. Still the success of these endeavours will rely on peoples' willingness and ability to be able to take risks. There has been discussion of how some emerging disruptive and innovative companies within the luxury industry embrace values in a more fundamental way (Bendell 2012). Entrepreneurs must undertake fundamental transformations if they activities will become deemed as sustainable (Egri and Herman 2000). It is therefore important to support active participation of Indigenous people as entrepreneurs who actively participate in securing investment for the implementation and evaluation of Indigenous Tourism Nation-Building Projects.

An example of Indigenous Tourism Nation-Building Projects is the World Indigenous Tourism Alliance (WINTA) which was formed in 2012 by the collective action of Indigenous tourism organisations from Australia, Canada, Nepal, New Zealand, Sweden and the USA. The formation of WINTA, in turn, supported a global dialogue in Darwin, Australia when 191 delegates from 16 countries representing Indigenous communities, the tourism industry, government agencies and other supporting bodies, came together at the Pacific Asia Indigenous Tourism Conference 2012, to commit to the development of Indigenous Sustainable Tourism and the promulgation of the Larrakia Declaration.

WINTA continues to evolve and to develop as an Indigenous-led global tourism network of Indigenous and non-Indigenous peoples and organisations. WINTA's objective is to collaborate with Indigenous communities, tourism industry entities, states and Non- Governmental Organisations which have an interest in addressing the aspirations of Indigenous peoples seeking empowerment through tourism and producing mutually beneficial outcomes (www.winta.org/). The Larrakia Declaration was declared to give practical expression to the United Nations Declaration on the Rights of Indigenous Peoples through Tourism. The conference delegates resolved to adopt the Declaration as the principles to guide international policy and better practice to encourage higher rates of tourism among local Indigenous, Native peoples and First Nations to showcase and ground sustainable development of Indigenous tourism across the globe (WINTA 2012). The principles of the Larrakia Declaration were subsequently recognised and supported by the UN World Tourism Organisation (UNWTO) which recognised the role of WINTA to facilitate, advocate and network with each affiliated Indigenous tourism body and with industry, governments and multilateral agencies.

WINTA builds upon US evidence which suggests that Native/Indigenous nations can progress towards their goals through 'nation building'—that is, by exercising genuine decision-making control over their affairs, creating or reinvigorating effective and legitimate institutions of self-government, setting strategic direction and developing public-spirited leadership (Jorgensen 2007; Harvard Project on American Indian Economic Development 2008). Evidence from a growing number of other settings—including Australia (Hunt et al. 2008; Hemming and Rigney 2008), Canada (Peeling 2004), and New Zealand (Goodall 2005), hints at the broader applicability of the results in transitioning Indigenous peoples from poverty to wealth creation. This also highlights the importance of developing education, vocational training and business skills (Altman and Finlayson 2003; Dyer et al. 2003; Fuller et al. 2005; Zeppel 2001).

There needs to be an awareness that if Indigenous tourism should succeed, it will be dependent on a new policy context to support innovative indigenous entrepreneurs. Fletcher et al. (2016) argue that this could only happen when both Indigenous stakeholders and increased diversity within Indigenous tourism product development are supported. Sustainable luxury tourism must be one part of that increased diversity.

6   Concluding Thoughts

It is the challenge and the opportunity for Indigenous groups around the Mardoowarra to develop both its management of nature and its model of governance. Luxury tourism via Indigenous governance could well be the disruptive innovation to counter unsustainable mass-consumption and a reason why international tourists will support these projects. Sustainable luxury tourism on Indigenous land will work best if it is achieved through a focus on small and community-based sustainable luxury tourism aimed for overseas visitors, where people can enjoy bespoke and unique access to an unparalleled wilderness. One of the most important aspects is to highlight the Indigenous values and the actual experience of these in a post and ongoing Native Title era.

Kleanthous argues that luxury is becoming more of a way of expressing deeper values which means becoming less wasteful. He notes that "sustainable luxury is a return to the essence of luxury with its ancestral meaning, to the thoughtful purchase, to the artisan manufacturing, to the beauty of materials in its broadest sense and to the respect for social and environmental issues" (cited in Gardetti 2014: 32).

Arguments to enhance the claim that Indigenous sustainable luxury tourism could support a return to original roots of luxury, and create unique, spiritual experiences for tourists in an unparalleled vast and pristine landscape are many. That it would occur at the invitation and supervision of the traditional custodians of an area such as the Mardoowarra, adds a rich dimension to such and enlightened possibility.

Acknowledgements This research is supported by Australian Government Research Training Program (RTP) Scholarship

We would also like to thank Johnny Edmonds from World Indigenous Tourism Alliance for helpful improvements.

References

Aboriginal and Torres Strait Islander Affairs (2003) Guide to provide advice for Indigenous tourism business—Media release. Canberra: Minister for Immigration and Multicultural and Indigenous Affairs (http://www.atsia.gov.au/media/ruddock_media03/r03064.htm). Accessed 8 Sept 2004

Agyeman J, Evans B (2004) Just sustainability: the emerging discourse of environmental justice in Britain? Geogr J 170(2):155–164

Altman J (1993) Indigenous Australians in the National Tourism Strategy: Impact. Sustainability and Policy Issues. Centre for Aboriginal Economic Policy Research. ANU, Canberra

Altman JC, Finlayson J (1992) Aborigines, tourism and sustainable development. CAEPR Discussion Paper No. 26, Centre for Aboriginal Economic Policy Research, ANU, Canberra

Altman J, Finlayson J (2003) Aborigines, tourism and sustainable development. J Tourism Stud 14 (1):78–91

Andersen J, Larsen JE, Møller IH (2009) The exclusion and marginalisation of immigrants in the Danish welfare society: dilemmas and challenges. Int J Sociol Soc Policy 29(5/6):274–286

De Barnier V, Falcy S, Valette-Florence P (2012) Do consumers perceive three levels of luxury? A comparison of accessible, intermediate and inaccessible luxury brands. J Brand Manage 19 (7):623–636

Bendell J, Kleanthous A (2007) Deeper Luxury. WWF, London

Bendell J (2012) Elegant Disruption: How Luxury and Society Can Change Each-Other for Good. Griffith University, Queensland

Bennett N, Lemelin RH, Koster R, Budke I (2012) A capital assets framework for appraising and building capacity for tourism development in aboriginal protected area gateway communities. Tour Manag 33:752766

Berry CJ (1994) The Idea of Luxury: A Conceptual and Historical Investigation. Cambridge University Press, New York

Bunten AC (2010) More like ourselves: indigenous capitalism through tourism. Am Indian Q 34 (3):285–311

Buultjens J, Waller I, Graham S, Carson D (2002) Public sector initiatives for aboriginal small business development in tourism. Centre for Regional Tourism Research Occasional Paper (No. 6) in partnership with Aboriginal Tourism Australia, Southern Cross University, Lismore

Buultjens J, Waller I, Graham S, Carson D (2005) Public sector initiatives for aboriginal small business development in tourism. In: Ryan C, Aicken M (eds) Indigenous tourism: the commodification and management of culture. Elsevier, Oxford, pp 127–147

Casula Vifell Å, Thedvall R (2012) Organizing for social sus-tainability: governance through bureaucratization in meta-organizations. Sustain Sci Practice Policy 8(1)

Christensen CM (1997) The innovator's dilemma: when new technologies cause great firms to fail. Harvard Business Press, Boston

Christensen CM, Raynor ME (2003) The innovator's solution: creating and sustaining successful growth. Harvard Business Press, Boston

Christensen CM, Overdorf M (2004) Meeting the challenge of disruptive change. In: Burgelman RA, Christensen CM, Wheelwright SC (eds) Strategic management of technology and innovation, 4th edn. McGraw-Hill Irwin, New York, pp 541–549

Colton JW, Whitney-Squire K (2010) Exploring the relationship between aboriginal tourism and community development. Leisure/Loisir 34(3):261278

Commonwealth of Australia (2003) Tourism White paper: a medium to long-term strategy for tourism. Department of Industry, Science and Resources, Canberra

Commonwealth of Australia (2010). Indigenous Economic Development Strategy—Draft for Consultation. http://resources.fahcsia.gov.au/IEDS/ieds_strategy_v4.pdf

Commonwealth of Australia (2011) Gazette No S132, Wednesday, 31 August 2011,https://www. environment.gov.au/heritage/laws/publicdocuments/pubs/106063_gazette_place_inclusion_20110831.pdf

Coria J, Calfucura E (2012) Ecotourism and the development of indigenous communities: the good, the bad, and the ugly. Ecol Econ 73:47–55

Davidson M (2009) Social sustainability: a potential for politics? Local Environ 14(7):607–619

Dempsey N, Bramley G, Power S, Brown C (2011) The social dimension of sustainable development: defining urban social sustainability. Sustain Dev 19(5):289–300

Department of Industry, Tourism & Resources [Tourism Division. Commonwealth of Australia]. Department of Industry, Tourism and Resources (2005). Request for tender: Research project —Business ready program for Indigenous tourism. Canberra

Dillard J, Dujon V, King M (eds) (2009) Understanding the Social Dimension of Sustainability. Routledge, New York

Dixon SEA, Clifford A (2007) Ecopreneurship: a new approach to managing the triple bottom line. J Org Change Manage 20(3):32645

Donaldson, J. (2007, June) Development and Poverty Reduction in Guizhou and Yunnan. China Q 190: 333–351

Dyer P, Aberdeen L, Schuler S (2003) Tourism impacts on an Australian Indigenous community: a Djabugay case study. Tour Manag 24(1):83–95. doi:10.1016/S0261-5177(02)00049-3

Egri CP, Herman S (2000) Leadership in North American environmental sector: values leadership styles and contexts of environmental leaders and their organizations. Acad Manage J 43 (4):523–553

Finlayson J, Madden R (1995) Regional tourism case studies: indigenous participation in tourism in Victoria. In: Bureau of Tourism Research (ed.), Tourism research and education in Australia. Proceedings from the Tourism and Education Conferences, Queensland, BTR, Canberra, (pp. 269–275)

Fiszbein A (1997) Lessons from Columbia. World Dev 25(7):1029–1043

Fitzroy River Declaration (2016) at http://www.klc.org.au/news-media/newsroom/news-detail/2016/11/15/kimberley-traditional-owners-unite-for-the-fitzroy-river

Fletcher C, Pforr C, Brueckner M (2016) Factors influencing Indigenous engagement in tourism development: an international perspective. J Sustain Tour 24(8–9):1100–1120

Freire P (2005) Pedagogy of the oppressed, 30th anniversary edn. Continuum, New York

Fuller D, Antella J, Cummings E, Scales J, Simon T (2001) Ngukurr field report: Stage two. Research Report No. 5, Flinders University, Adelaide

Fuller D, Buultjens J, Cummings E (2005) Ecotourism and Indigenous microenterprise formation in northern Australia opportunities and constraints. Tour Manag 26(6):891–904

Gardetti MA (2011) Sustainable Luxury in Latin America, conference dictated at the Seminar Sustainable Luxury & Design, Instituto de Empresa (Business School), Madrid, Spain

Gardetti MA (2014) Stories from the social pioneers in the sustainable luxury sector: a conceptual vision, Sustainable Luxury and Social Entrepreneurship, edited by María Eugenia Girón, and Miguel Angel Gardetti, Greenleaf Publishing

Girón ME (2012) Diccionario LID sobre Lujo y Responsabilidad. Editorial LID, Madrid

Goodall A (2005) Tai timu, tai pari... the ever-changing tide of Indigenous rights in Aotearoa New Zealand. In: Garth Cant, Anake Goodall, and Justine Inns, (eds), Discourses and Silences: Indigenous Peoples, Risks and Resistance. Christchurch: Department of Geography, University of Canterbury. pp. 185–98

Harvard Project on American Indian Economic Development (2008) The State of the Native Nations: conditions under U.S. Policies of Self-Determination. New York, Oxford University Press

Harvey D (1996) Justice, nature and the geography of difference. Blackwell

Heine K (2011) The concept of luxury brands. Technische Universität Berlin, Department of Marketing, Berlin

Hemming S, Daryle R (2008) Unsettling Sustainability: Ngarrindjeri political literacies, strategies of engagement and transformation'. Continuum J Media Cultural Stud 22(6):757–775

Higgins-Desbiolles F, Trevorrow G, Sparrow S (2014) The Coorong Wilderness Lodge: a case study of planning failures in Indigenous tourism. Tour Manag 44:4657

Hinch T, Butler R (2007) Introduction: revisiting common ground. In: Butler R, Hinch T (eds) Tourism and Indigenous peoples: Issues and implications. Butterworth-Heinemann, Burlington, MA, pp 1–13

Hockerts K, Wüstenhagen R (2009) Emerging Davids versus Greening Goliaths (Frederiksberg. Denmark, CBS Center for Corporate Social Responsibility)

Hunt J, Smith D, Garling S, Sanders W (eds) (2008) Contested governance: culture, power and institutions in indigenous Australia. Australian National University, Canberra

Ivory B (2003) Poverty and enterprise. In Carr SC, Sloane TS (eds), Poverty and psychology: emergent critical practice. Kluwer Academic/Plenum Publishers, New York, USA

Jorgensen A-M (2007) Rebuilding native nations, strategies for governance and development. The University of Arizona Press, Tuscon, Arizona

Kapferer J-N (2010, November–December) All that glitters is not green: the challenge of sustainable luxury. Europ Business Rev pp. 40–45

Kapferer J-N, Michaut A (2015) Luxury and sustainability: a common future? The match depends on how consumers define luxury. Lux Res J 1(1):3–17

Kearins K, Collins E, Tregidga H (2010) Beyond corporate environmental management to a consideration of nature in visionary small enterprise. Business Soc 49:512

Langton M, Rhea ZM, Palmer L (2005) Community oriented protected areas for indigenous peoples and local communities. J Political Ecolo 12:23–50

Littig B, Grießler E (2005) Social sustainability: a catchword between political pragmatism and social theory. Int J Sustain Dev 8(1–2):65–79

Ludes, P (1997) Aufstieg und Niedergang von Stars als Teilprozess der Menschheitsentwicklung i W. Faulstich, och H. Korte (red) Der Star Geschichte Rezeption Bedeutung, München: Wilhelm Fink Verlag

Mair J, Laing JH (2013) Encouraging pro-environmental behaviour: the role of sustainability-focused events. J Sustainable Tour 21(8):0001

Marcuse P (1998) Sustainability is not enough. Environ Urbanizat 10(2):103–111

Mary River Statement https://www.nailsma.org.au/hub/resources/.../mary-river-statement-2009.html

Miller G, Rathouse K, Scarles C, Holmes K, Tribe J (2010) Public understanding of sustainable tourism. Ann Tour Res 37(3): 627–645

Mowforth M, Munt I (1998) Tourism and sustainability: new tourism in the third World. Routledge. London

Ockwell D, Mallett A (2013) Low carbon innovation and technology transfer. In: Urban F, Nordensvard J (eds) Low carbon development: key issues. Routledge, Abingdon, pp 109–128

Peeling AC (2004) Traditional Governance and Constitution Making among the Gitanyow. Prepared for the First Nations Governance Centre. Available at http://fngovernance.org/resources_docs/Constitution_Making_Among_the_Gitanyow.pdf

Pepper M, Keogh S (2014, August) Biogeography of the Kimberley, Western Australia: a review of landscape evolution and biotic response in an ancient refugium. J Biogeograp 41(8) (): 1449

Pink B, Allbon P (2008) The health and welfare of Australia's Aboriginal and Torres Strait Islander Peoples 2008. Australian Bureau of Statistics & Australian Institute of Health and Welfare, Canberra

Poelina A, McDuffie M, (2016a), Our Shared and Common Future. Available at https://vimeo. com/132022090. Password: Kimberley, Madjulla Association, Broome, Western Australia

Poelina A, McDuffie M, (2016b) Mardoowarra's Right to Life. Available at https://vimeo.com/ 205996720. Password: Kimberley, Madjulla Association, Broome, Western Australia

Ryan C, Huyton J (2002) Tourists and aboriginal people. Ann Tour Res 29(3):631–647

Schlosberg D (2013) Theorising environmental justice: the expanding sphere of a discourse. Environ Politics 22(1):37–55

Schlosberg D, Carruthers D (2010) Indigenous struggles, environmental justice, and community capabilities. Global Environ Politics 10(4):12–35

Schmiechen J, Boyle A (2007) Aboriginal tourism research in Australia. In: Butler R, Hinch T (eds) Tourism and Indigenous peoples: issues and implications. Butterworth-Heinemann, Burlington, MA, pp 58–72

SCRGSP (Steering Committee for the Review of Government Service Provision) (2016) Overcoming indigenous disadvantage: key indicators 2016. Productivity Commission, Canberra

Semeniuk V, Brocx M (2011) King Sound and the tide-dominated delta of the Fitzroy River: their geoheritage values. J Royal Soc Western Australia 94:151–160

Steering Committee for the Review of Government Service Provision (SCRGSP) (2016) Overcoming Indigenous Disadvantage: Key Indicators. Productivity Commission, Canberra

Stamm A, Dantas E, Fischer D, Ganguly S, Rennkamp B (2009) Sustainability-oriented innovation systems: towards decoupling economic growth from environmental pressures? Discussion paper 20/2009. Bonn, German Development Institute

Tran Tran (2015) Chapter 8:|The non-legal guide to meaningful recognition: a case study from the Canning Basin, West Kimberley. In: Sillitoe Paul (ed) Indigenous studies and engaged anthropology: The collaborative moment. Routledge, Abigdon

Tourism Research Australia (2010) Indigenous tourism in Australia: profiling the domestic market. Canberra: tourism research Australia. http://www.ret.gov.au/tourism/

Turner KL, Berkes F, Turner NJ (2012) Indigenous perspectives on ecotourism development: a British Columbia case study. J Enter Commun People Places Global Economy 6(3):213–229

United Nations (2017) United Nations permanent forum on indigenous issues background guide, online at https://www.humanrights.gov.au/united-nations-permanent-forum-Indigenous-issues

United Nations (2017) United Nations declaration on the rights of Indigenous Peoples at http://www.un.org/esa/socdev/unpfii/documents/DRIPS_en.pdf

Urban F, Nordensvärd J, Zhou Y (2012) Key actors and their motives for wind energy innovation in China. Innovation and Development 2(1):111–130

Vogwill R (2015) Water Resources of the Mardoowarra (Fitzroy River) Catchment. The Wilderness Society, Perth, Western Australia

Walker K, Moscardo G (2016) Moving beyond sense of place to care of place: the role of Indigenous values and interpretation in promoting transformative change in tourists' place images and personal values. Journal of Sustainable Tourism 24(89):0001

Watson J, Watson A, Poelina A, Poelina N, Watson W, Camilleri J, Vernes T, (2011) Nyikina and Mangala Mardoowarra Wila Booroo: Natural and Cultural Heritage Plan. Nyikina-Mangala Aboriginal Corporation and World Wide Fund for Nature—Australia

Weaver D (2009) Indigenous tourism stages and their implications for sustainability. J Sustain Tourism 18(1):43–60

Wells K (2015) Australia's wild rivers. Australian Government: http://www.australia.gov.au/ about-australia/australian-story/australias-wild-rivers

Western Australian Aboriginal Heritage Act (1972) State Law Publisher: http://www.slp.wa.gov. au/PDF

West Kimberley (2011) Western Australia, Australian National Heritage Place, https://www. environmenta.gov.au/heritage places/national, 2011/west kimberley

West Kimberley National Heritage Place—A draft guide for landholders (2012) at https://www. environment.gov.au/system/files/pages/ed0b4e39.../landholders-guide

WINTA (2012) The Larrakia Declaration, Accessed 19th March, http://www.winta.org/wp-content/uploads/2012/08/The-Larrakia-Declaration.pdf

Whitford M, Ruhanen L (2009) Indigenous tourism businesses in Queensland: criteria for success. Sustainable Tourism Co-Operative Research Centre, Gold Coast, Qld

Whitford MM, Ruhanen LM (2010) Australian Indigenous tourism policy: practical and sustainable policies. J Sustain Tourism 18(4):475–496

World Commission on Environment and Development (1987) Our Common Future. Oxford University Press, Oxford

Young E (1988) Aboriginal economic enterprises: problems and prospects. In Wade-Marshall D, Loveday P (eds), Contemporary issues in development: north australia: progress and prospects. (vol 1, pp. 182–200). NARU Darwin

Zeppel H (2001) Aboriginal cultures and Indigenous tourism. In: Douglas N, Douglas N, Derrett R (eds) Special interest tourism. Wiley, Milton, Queensland, pp 232–259

Design Similarity as a Tool for Sustainable New Luxury Product Adoption: The Role of Luxury Brand Knowledge and Product Ephemerality

Feray Adıgüzel, Matteo De Angelis and Cesare Amatulli

Abstract Despite rising demand for brands that can combine luxury with the zeal to make the world better, many luxury brands still hesitate to introduce environmentally (or green) sustainable new luxury products. However, indulgence without guilt is possible by correctly marketing sustainable new luxury products. Even more, introducing sustainable new luxury products can be an effective way for luxury companies to differentiate their offering and get competitive advantage. We argue that luxury and sustainability are not incompatible concepts when luxury brands do employ the right product design strategy. Indeed, the effectiveness of two new green luxury product design strategies have been investigated in depth in this chapter. First, the green new product might be similar in design to luxury company's previous non-green products. Second, the green new product might be similar in design to models of a different, non-luxury company specialized in green production. We investigate the effect of design strategy on new product purchase intention, and propose that such an effect might be affected by fit (i.e. moderated mediated) by the combination of one product-related factor, such as product ephemerality (i.e., how long lasting the new product is), and one consumer-related factor, such as consumers' luxury brand knowledge (i.e., how much consumers know about the brand and its past). We empirically test our account through an experimental study, whose results show that lowly and medium knowledgeable consumers might react negatively to a new green product similar to green company's models due to reduced fit perceptions when the new green luxury product is ephemeral. Such an effect was not observed for durable new luxury products.

F. Adıgüzel (✉) · M. De Angelis
Department of Business and Management, LUISS Guido Carli University,
Viale Romania, 32, 00197 Rome, Italy
e-mail: fadiguzel@luiss.it

M. De Angelis
e-mail: mdeangelis@luiss.it

C. Amatulli
Ionian Department of Law, Economics and Environment, University of Bari,
Via Duomo, 259, 74123 Taranto, Italy
e-mail: cesare.amatulli@uniba.it

© Springer Nature Singapore Pte Ltd. 2018
M. A. Gardetti and S. S. Muthu (eds.), *Sustainable Luxury, Entrepreneurship, and Innovation*, Environmental Footprints and Eco-design of Products and Processes, https://doi.org/10.1007/978-981-10-6716-7_9

Keywords Luxury brands · Sustainability · Design · New products
Design similarity · Ephemerality · Brand knowledge

1   Introduction

In many sectors, sustainability is considered one of the most prominent drivers of companies' innovation and growth (e.g., Cho et al. 2013; Lloret 2016; Lozano and Huisingh 2011). Interestingly, the pressure to devote more attention to sustainability initiatives has increased in the last decade in the luxury sector (Davies et al. 2012; Kapferer and Michaut-Denizeau 2014; Janssen et al. 2014). In particular, while not traditionally involved in pro-environmental actions, luxury companies have, in recent times, significantly strengthened their commitment toward reducing their environmental impact. For instance, Stella McCartney, one of the leading luxury fashion brands, introduced in 2010 some innovative shoes made with biological components thought to replace leather. More recently, Gucci launched an innovative model of sunglasses made with liquid wood, a bio-degradable, eco-friendly material alternative to plastic (http://www.kering.com/en/sustainability/achievements).

We focus on consumers' perceptions of sustainable product innovation by investigating the issue of luxury companies introducing environmentally sustainable (hereafter, green) new products—that is, those products manufactured with attention to minimizing the exploitation of natural resources, the use of toxic materials, or the emission of waste and pollutants (Haws et al. 2014; Lin and Chang 2012). Our aim is to understand what actions companies might undertake to increase consumers' attitudes toward and intentions to purchase new green luxury products (hereafter, NGLP). Similar to previous literature (e.g., Griskevicius et al. 2010; Lin and Chang 2012; Olsen et al. 2014) throughout this chapter, we use the term "green products" interchangeably with the term "environmentally sustainable products" referring to products that offer environmentally sustainable features. Past research into this issue has investigated the role of factors such as the marketing communication tactics (e.g., Kronrod et al. 2012) and consumers' motivations and characteristics as drivers of consumers' attitudes and intention to purchase green products (e.g., Griskevicius et al. 2010; Schlegelmilch et al. 1996). By contrast, we explore how consumers' purchase intentions are affected by the type of new green product introduced. Specifically, we focus on the impact of the new green product's aesthetic design (Cucuzzella 2016; Fletcher 2013).

Amidst luxury companies' growing interest in environmental sustainability, academia has devoted relatively little attention to understanding whether and under what conditions luxury consumers might be willing to embrace sustainable luxury products (Boenigk and Schuchardt 2013; Steinhart et al. 2013). Scholarly research has typically investigated the issue of sustainable consumption in terms of non-luxury products such as food or cosmetics (Johnston et al. 2011; Ngobo 2011). Granted, this lack of research may stem from the fact that luxury and sustainability are still generally seen as incompatible (e.g., Beckham and Voyer 2014; Kapferer and Michaut-Denizeau 2014; Torelli et al. 2012), even though the two concepts share some essential elements such as durability, rarity and beauty (Janssen et al. 2014; Kapferer 2010). Likewise, despite some evidence that luxury consumers are increasingly aware of the environmental consequences of luxury manufacturing practices (see, for instance, Achabou and Dekhili 2013), most research to date indicates that sustainability is a relatively unimportant driver of their purchasing behavior (Davies et al. 2012; Griskevicius et al. 2010)—if not simply deleterious to consumers' perceptions about the quality of luxury items (Achabou and Dekhili 2013).

In this chapter, we investigate the conditions that make luxury consumers more or less likely to accept such NGLP. Indeed, recent research in non-luxury contexts (Olsen et al. 2014) has demonstrated that companies generally benefit from the introduction of green new products in terms of consumers' purchase intentions (Sen and Bhattacharya 2001) and referrals (Du et al. 2007). In following this line, we test the differential effectiveness of two alternative strategies that luxury companies might adopt when introducing green new products, whereby such products vary on the basis of drivers that are crucial for luxury consumers' purchasing decisions, such as aesthetics and design (Fuchs et al. 2013).

Importantly, on the one hand a luxury company may want to preserve its stylistic identity and leverage the quality perception associated with its brand, thus it should introduce green new products that match the aesthetic characteristics of its own previous, non-green models. One example of this comes from Porsche, whose Panamera hybrid model parallels the company's traditional car models in terms of design. However, on the other hand, a luxury company may want to be perceived as highly innovative and committed to environmental sustainability, and thus may introduce a green new product that approximates a product marketed by another company that specializes in green production (hereafter, the green company). Case in point, Lexus recently launched a hybrid car, the Lexus CT 200H, which has been criticized for closely borrowing from the design of the Toyota Prius, a widely known model of green car. The question facing this research, then, is how does the adoption of one or the other strategy influence consumers' attitudes toward and intention to buy new green luxury products?

To answer this question, we conducted an experiment, aiming to manipulate the green product introduction strategy and investigate the general effect of each on consumers' luxury new green luxury product purchase intention. In our experiment, we especially tested the mediating role of fit perception—that is consumers' perception of congruity between the NGLP and the luxury brand introducing it—and the moderating role of both consumer-related and product-related factors simultaneously. In brief, our findings suggest that whether consumers prefer a NGLP that parallel the luxury manufacturer's previous models or a green new product that resembles a green company's models depends on relevant consumer-related and product-related characteristics. In particular, our empirical analysis demonstrates that intention to purchase NGLP resembling a green company's models decreases when consumers have low or average values of brand knowledge and, simultaneously, the product in question is ephemeral (rather than durable) in nature, that is with a long-term orientation, tending to go slowly in and out of trends. In addition, we demonstrated such an effect is mediated by consumers' perceptions of fit between the NGLP and the luxury brand image.

Overall, this research makes a meaningful contribution to academic research on sustainable luxury marketing and consumption by demonstrating that consumers' reactions to a luxury green new product crucially depends on the type of product, the product's design and manufacturing characteristics, and customer-related factors. Our research might prove insightful for luxury marketing managers seeking to increase consumers' acceptance of green new products, thus fostering a greater convergence between luxury and sustainability.

2 Sustainability and Green Products

Environmental and social problems have greatly evolved, due to the rapid development activities by humans, especially since the beginning of the industrialization era. These problems, namely global warming, ozone depletion, water and air pollution, loss of species, and farmland erosion, even threatens the environment and human life nowadays. Overconsumption of natural resources, especially in the industrial nations, call for actions aimed at securing sustainable development and consumption.

Sustainable development and consumption have been common interest in last thirty years especially since the publication of the Brundtland Report, titled "Our Common Future" in 1987 by the World Commission on Environment and Development (WCED 1987). Sustainable consumption and production were defined as "the use of services and related products, which respond to basic needs and bring a better quality of life while minimizing the use of natural resources and toxic materials as well as the emissions of waste and pollutants over the life cycle of the service or product so as not to jeopardize the needs of further generations" (Peattie and Peattie 2009). Another definition of sustainable consumption is "the consumption that supports the ability of current and future generations to meet their material and other needs, without causing irreversible damage to the environment or loss of function in natural systems" by the Oxford Commission on Sustainable Consumption (OCSC 2000). The concern about environmental and ethical issues amongst consumers have started to influence their behavior and their lifestyles (Vandermerwe and Oliff 1990; Worcester 1993). More importantly, consumers expect sustainable product introductions from companies. Consumer expectations, legal, market and financial pressures act as an impulse for innovation and the development of new sustainable products. Ethical and environmental performance as a new product attribute have become a source of potential differentiation and competitive advantage for companies (Porter and van der Linde 1995).

Ljungberg (2007) defined a sustainable product as a product that gives as little impact on the environment as possible during its life cycle. The life cycle of a product consists of extraction of raw material, production, use and final recycling (or deposition). We must point that the impact on environment cannot be zero, but it must be convincingly reduced. Very good overview of different definitions of green products in literature and classification of green products can be found in Dangelico and Pontrandolfo (2010). Such scholars categorized product life cycle stages into three phases, namely (1) before usage (including materials' extraction, production processes, transportation processes), (2) usage, and (3) after usage (end-of-life), thereby distinguishing green products on the basis of their main environmental focus. In their framework, green products could be those focused on materials, those focused on energy, and those focused on pollution/toxic waste. While sustainable product definition covers the whole life cycle of a product, green products could be related to at least one lifecycle phase alone. If a new product is green, it is also sustainable because it has less environmental impact in comparison to a conventional non-green product. That is why, the term "green products" can be used interchangeably with the term "sustainable products".

3 Luxury and Sustainability

Luxury can be considered a crisis-proof sector (Smith 2009). Indeed, despite the global economic crisis, the luxury industry anticipates an estimated 9% annual growth rate at least until 2020 (Bain and Company 2014). Such growth is fueled by the exploding number of luxury consumers, who are expected to total 465 million by 2021, compared to 2014s 390 million (The Boston Consulting Group 2015). As global luxury consumers have steadily transitioned from the privileged few to the "happy many" (Dubois and Laurent 1998), their interest in and awareness of environmental issues has expanded (AFP 2008), which has prompted heightened expectations about the environmental dimension of business (Lochard and Murat 2011). These trends suggest luxury and sustainability are compatible concepts—especially when paired with the idea that the two share some elements, like quality, rarity, durability, and the preservation of ancestral skills, raw materials and local activities (Janssen et al. 2014; Kapferer 2010). In response to consumers' increasing awareness about the potential environmental effects of luxury manufacturing, some luxury companies are striving to improve their production processes and render their final products more sustainable. Hermès, for instance, advertises its long running commitment to recycling practices (http://www.telegraph.co.uk/luxury/womens-style/15811/herm%C3%A8s-embraces-recycling-with-petit-h.html), while Armani is committed to reducing its emission of hazardous chemicals in accordance with the "zero discharge goal" by 2020 (http://alive.armani.com/news/it/giorgio-armani-commitment-to-zero-discharge/).

As scholarly research indicates, however, these companies still face the hurdle of consumers perceiving luxury and sustainability as antagonistic concepts (e.g., Achabou and Dekhili 2013; Beckham and Voyer 2014; Kapferer and Michaut-Denizeau 2014; Strong 1997; Torelli et al. 2012). Indeed, luxury is often associated with excess, personal pleasure, superficiality and ostentation, while sustainability typically evokes altruism, sobriety, moderation and ethics (Widloecher 2010). Consistent with this idea, research suggests that environmental protection—and, by extension, the sustainability of luxury products and brands—is a factor of secondary importance for luxury consumers (Davies et al. 2012; Griskevicius et al. 2010). In fact, scholars have demonstrated that many consumers see luxury goods as having less short-term and long-term impact on the environment relative to non-luxury products (Nia and Zaichkowsky 2000; Vigneron and Johnson 2004; Ward and Chiari 2008). In some cases, a sustainability emphasis may even negatively affect consumers' overall perception of luxury goods' quality (Achabou and Dekhili 2013). Nonetheless, there is a need for more research that identifies how luxury companies' sustainability practices affect consumers' attitudes and behaviors toward green luxury products, which is what we address in the present chapter.

4 Innovation in Luxury Through New Green Products

In the last decade, companies' environmental policies have gradually shifted from adopting clean technologies and pollution prevention initiatives to producing green products (Albino et al 2009; Chung and Tsai 2007; Gershoff and Frels 2015; McKinsey 2011; Pujari 2006). This shift has largely been in response to consumers' generally positive attitudes toward green products and their producers (e.g., Du et al. 2007; Pickett-Baker and Ozaki 2008). Because of companies' rising interest in introducing green products to the market, there is a need to understand whether and how different types of green new products may engender different consumer responses. However, this issue has not been adequately addressed to date. Instead, previous studies on green products have focused on issues such as the diffusion of green products at the aggregate level (Janssen and Jager 2002), the marketing communication tactics aimed at increasing consumers' buying intentions (Kronrod et al. 2012; Yang et al. 2015), the attitudes and motivations underlying consumers' decision to buy green products (Griskevicius et al. 2010; Van Doom and Verhoef 2011), and the type of consumers who are more willing to buy such products (Schlegelmilch et al. 1996; Shrum et al. 1995). The research most relevant to our study was performed by Olsen et al. (2014), who assessed the factors that moderate the effect of green new product introductions on brand attitude, such as the product's category (i.e., virtue vs. vice product categories), the framing of the communication, and the company's level of environmental legitimacy. However, their study focused squarely on fast-moving consumer goods in non-luxury categories.

Our study fills this gap by studying the effect of green new product design on consumers' attitudes and behavior in a sector, such as luxury, where aesthetics and design are key drivers of consumer decisions. In particular, we investigate whether and under what conditions consumers are more likely to prefer a green new product that resembles the previous models of either the luxury company in question or a well-known green company. In the next section, we develop our hypothesis regarding the expected outcome.

5 Luxury Brand Knowledge and Product Ephemerality

Consumers seem to perceive lower overall performance of green products compared to conventional alternatives (see, for instance McKinsey and Company 2008) and luxury consumers are even more sceptical about the quality of green items (e.g., Achabou and Dekhili 2013; Griskevicius et al. 2010; Magnoni and Roux 2012). Indeed, building on the well-established bookkeeping model of information processing (Gürhan-Canli and Maheswaran 1998; Loken and John 1993; Milberg et al. 1997), we argue that when a luxury brand introduces a new green product, because of the resulting incongruency between such a new product and the schema (i.e., the luxury brand), consumers' product evaluation and purchase intention might decrease. In other words, the expected negative effect of the luxury brands' decision to introduce new green products would be due to a decrease in consumers' perceptions of fit between the new item and the schema. In contrast, for luxury brands, introducing new products that are similar in design to their existing and established models might be a better strategy, because it would allow luxury brand to maintain a certain level of consistency with their past, their heritage, and their stylistic identity. These arguments lead to predict that when luxury brands introduce green new products that aesthetically resemble their own established models, consumers could be less likely to lower their brand evaluation and purchase intention compared to when they introduce green new products that aesthetically resemble models of a green company.

We propose, however, that while plausible considering the nature of luxury goods and luxury brands, this negative effect of introducing green new products resembling a green company's models may be importantly shaped by consumers' familiarity with the luxury brand in question, i.e., with how knowledgeable about the brand, its past, and its stylistic identity consumers are. Indeed, consumers vary in their knowledge of luxury fashion brands. Interestingly, brand knowledge can help shape consumer brand evaluations and preferences (Esch et al. 2006). In particular, it seems that consumers with higher brand knowledge have stronger brand preference than consumers with lower brand knowledge (Kirmani et al. 1999; Sujan and Dekleva 1987). This tendency should be magnified for luxury brands because of the emotional attachment that typically characterizes consumers' relationship with status-signaling products (Atwal and Williams 2009). Moreover, consumers with a high level of brand knowledge are particularly concerned about the status signaled by that brand, which might be undermined by the brand's association with a green, non-luxury brand (Fuchs et al. 2013; Han et al. 2013). Consistent with this idea, Kirmani et al. (1999) have demonstrated that brand ownership, which is positively correlated with brand knowledge, leads consumers to react negatively to a downward association.

It is also possible that, in addition to consumer-related factors, also the intrinsic characteristics of luxury products might play a role in consumers' assessment of NGLP. In particular, in the context of luxury and sustainability, consumers' assessments of new products may depend on said products' relative level of ephemerality versus durability (e.g., Berthon et al. 2009; Janssen et al. 2014). Ephemeral products, such as clothing, are those with a short-term orientation, tending to go quickly in and out of trends. By contrast, durable products, such as watches, are typically more enduring and wearable for more than one season (Kapferer 1998; Stock and Balachander 2005). Durability is considered one of the main characteristics of luxury products in general, as they must generally last longer than conventional ones in terms of style and use (Kapferer 2010), thus fitting with the long-term orientation of a CSR agenda (Janssen et al. 2014). Indeed, luxury products are usually characterized as being relatively classic and enduring in their ability to represent the style and design codes of the parent brand.

Building on the stated importance of consider consumer- and product-related factors as important drivers of consumers' assessment of NGLP resembling luxury company's versus a green company's models, here as follows, we present an empirical experimental analysis testing the joint effect of NGLP type, consumers' luxury brand knowledge and product ephemerality on consumers' purchase intentions.

6 Methodology and Results

6.1 Experimental Procedure

In order to analyze the simultaneous effect of product ephemerality and consumers' knowledge about the luxury brand on the introduction of NGLPs that are similar to green company's models or to previous established models of the luxury brand, an experimental approach was adopted. More specifically, the goal of the experiment was to understand whether and how the effect of NGLP design similarity on consumers' purchase intention changes through fit perceptions when the level of consumers' knowledge about the luxury brand in question and the product ephemerality vary simultaneously. We investigated the effect of luxury brand knowledge and product ephemerality simultaneously and joint with the effect of design similarity on purchase intention through fit perceptions (which acted as mediating variable M). We employed Model 11 of the PROCESS macro for SPSS (Hayes and Preacher 2014; Preacher et al. 2007, see Fig. 1).

Design similarity and product type are nonmetric variables while luxury brand knowledge, NGLP-luxury brand fit, and purchase intention are metric variables. Hayes and Preacher's (2014) approach was followed for statistical moderated mediation analysis (Hayes 2013; Preacher et al. 2007), using the SPSS PROCESS macro to quantify direct and indirect effects involving a categorical variable (such

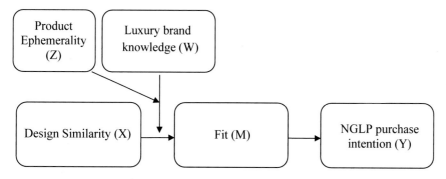

Fig. 1 Moderated mediation model with luxury brand knowledge and product ephemerality as moderators, fit as a mediator and new green luxury product (NGLP) purchase intention as a dependent variable

as design similarity). This macro uses bias-corrected bootstraps or Monte Carlo confidence intervals to test the indirect effects of a categorical independent variable. The bias-corrected bootstraps for indirect effects were computed considering that they should be considered significant if 0 falls outside the confidence interval (CI). Because brand familiarity is continuous and product ephemerality is dichotomous, PROCESS automatically produces conditional effects of design similarity for each group among people relatively low familiar (1SD below the mean), moderate in brand familiarity (the mean) and relatively older (1SD above the mean) (Hayes 2013).

Giorgio Armani was selected as a luxury brand. Product ephemerality was manipulated by designing scenarios that described either a jacket (which is more ephemeral in nature) or a wallet (which is more durable in nature). As luxury brand Armani was used, and it was described as a leading luxury brand presenting either a new wallet model that is very long-lasting (in the durable condition) or a new haute couture jacket that is quite seasonal (in the ephemeral condition). As green brands, two brands were used, one for the wallet and one for jacket category. The wallet brand was named a GreenFrames and it was described as one of the few wallet brands with a core focus on environmental sustainability, while the jacket brand was named GreenFashion and it was described as one of the few fashion brands with a core focus environmental sustainability.

One hundred and ninety-two U.S. participants (77 male, 115 female, M_{age} = 37 years, SD_{age} = 12.56, 11% with low monthly income (below $1000), 67% with average income [between $1000 and $5000), and 22% with high income (above $5000)] were recruited from Amazon.com's Mechanical Turk service and paid for their participation. They were randomly assigned to one of four conditions in a 2 (design similarity: similar to Armani's previous models vs. similar to GreenFrames'/GreenFashion's models) × 2 (product ephemerality: durable vs.

ephemeral) between-subjects experiment. Sample size of conditions varied between 45 and 50.

Design similarity was manipulated by telling participants that the luxury brand Armani had recently launched a new green model of either a jacket (ephemeral condition) or a wallet (durable condition), which looked similar in the design to either Armani's previous models or GreenFrames'/GreenFashion's models (see Appendix). GreenFrames'/GreenFashion's was described as one of the few fashion brands with a core focus environmental sustainability.

Participants did not see pictures of the new green product to avoid confounding effects related to how they interpret "aesthetic" of products, which might have influenced their brand evaluation and purchase intention (Creusen and Schoormans 2005).

Participants rated their overall assessment about the Armani brand on a four-item, seven-point semantic differential scale (negative/positive, unfavorable/favorable, bad/good, dislike/like) and purchase intention of a new product using a three-item scale adapted from Dodds et al. (1991) ("I would purchase the described Arman," "I would consider buying the described Armani," and "The probability that I would consider buying the described Armani is high;" 1 = totally disagree, 7 = totally agree). In order to check the validity of the ephemerality manipulation, participants were asked to rate how ephemeral versus durable they perceived the wallet and the jacket to be as product categories, using a five-item scale adapted from Janssen et al. (2014) (e.g., "This product can be worn for years and will never go out of fashion," "This product is worn and passed down from generation to generation;" 1 = totally disagree, 7 = totally agree). The fit between the NGLP and the luxury brand introducing it was measured on a two-item, seven-point semantic differential scale (dissimilar/similar, inconsistent/consistent; Martinez et al. 2008). Participants' general concern for the environment was measured using a six-item scale developed by Haws et al. (2014) (e.g., "It is important to me that the products I use do not harm the environment," "My purchase habits are affected by my concern for our environment" 1 = totally disagree, 7 = totally agree).

Ephemerality, fit and purchase intention. Measures of the perceived ephemerality (as a check of our manipulation, $\alpha = .82$), fit (i.e., the mediator; $\alpha = .92$), and NGLP purchase intention (i.e., the dependent variable; $\alpha = .96$) were reliable.

Control variables for internal validity. As control variables, concern for environment ($\alpha = .96$), prior luxury brand evaluation before scenarios ($\alpha = .97$) and income were statistically examined through ANOVA to inspect differences across four conditions. Results, however, showed that there were no significant differences in concern for environment [$F(3188) = 1.627$, n.s.], prior luxury brand evaluation [$F(3188) = 1.211$, n.s.] and income [$F(3188) = .502$, n.s.] across the four conditions.

7 Results

The check of the ephemerality manipulation confirmed that wallet was perceived as more durable, thus less ephemeral, than jacket [$M_{wallet} = 4.88$, $M_{jacket} = 3.24$, F (1190) = 87.102, $p < .001$]. Using a moderated mediation analysis though the SPSS PROCESS macro, a model in which fit was regressed on design similarity (coded as $0 =$ similar to Armani's previous models and $1 =$ similar to GreenFrames'/GreenFashion models), product ephemerality (coded as $0 =$ ephemeral and $1 =$ durable) and luxury brand knowledge as well as the two-way interactions and the three-way interaction among these factors was first conducted. Then, NGLP purchase intention (i.e., the dependent variable) was regressed on fit.

Of greatest importance to this investigation, the results showed a (marginally) significant three-way interaction effect on fit ($b = -.49$, $t = -1.75$, $p = .08$), a significant two-way interaction between design similarity and product ephemerality on fit ($b = 1.96$, $t = 2.25$, $p < .05$) and the main effect of design similarity on fit ($b = -1.06$, $t = -2.24$, $p = .08$). Among control variables, only prior brand evaluation influenced fit ($b = .30$, $t = 3.52$, $p < .001$), but concern for environment did not influence fit. Besides, there was a positive and significant effect of fit on NGLP purchase intention ($b = .29$, $t = 3.50$, $p < .001$).

The conditional effects of a three-way interaction is our interest. A closer inspection of such a three-way interaction was then obtained by analyzing at the conditional effects of design similarity on purchase intention via mediation of fit, which consists in looking at such an indirect effect of design similarity to the dependent variable at different values of luxury brand knowledge and product ephemerality (see Fig. 2). Such an analysis revealed a significant and negative indirect effect of design similarity on purchase intentions for ephemeral products and for consumers with either low (i.e., less than 1.15; $b = -.25$, 95% CI $= -.62$, $-.03$) or medium (i.e., $1.15 <$ luxury brand knowledge < 2.75; $b = -.18$, 95% CI $= -.42, -.01$) levels of luxury brand knowledge. In contrast, no evidence of significant indirect effects of design similarity on purchase intention was found for durable products at any level of luxury brand knowledge.

Overall, these results show that consumers with low and medium (but not those with high) luxury brand knowledge might perceive the luxury brand's decision to launch an ephemeral NGLP negatively when the NGLP is similar to green brand's models instead of luxury brands' established models due to reduce fit perceptions, while the data of this study did not show any significant variation in consumers' purchase intention associated with the launch of a durable NGLP. For durable products, results demonstrated that when the NGLP resembled the green company's models fit increased.

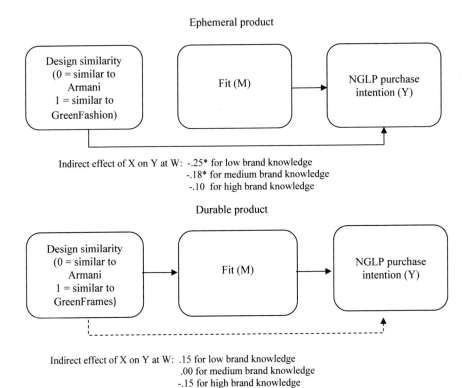

Ephemeral product

Indirect effect of X on Y at W: -.25* for low brand knowledge
-.18* for medium brand knowledge
-.10 for high brand knowledge

Durable product

Indirect effect of X on Y at W: .15 for low brand knowledge
.00 for medium brand knowledge
-.15 for high brand knowledge

Dotted arrows indicate non-significant indirect effects.
*significant at p < .05

Fig. 2 Indirect effects of design similarity on NGLP purchase intention among participants with different levels of brand knowledge via fit at the two levels of product ephemerality (W). *Dotted arrows* indicate non-significant indirect effects. *significant at p < .05

8   Conclusions

This research investigated consumers' attitudes toward the launch of sustainable new products, thus aligning sustainable consumption with sustainable innovation. Specifically, it explored whether and under what conditions companies' different sustainability-driven design strategies—particularly in highly design-based sectors such as luxury fashion and accessories—might positively influence purchase intentions toward NGLPs. As such, this research sheds light on what drives consumers to embrace products that are luxurious and green simultaneously.

Our experiment investigated the effect of NGLP's design similarity on purchase intentions considering simultaneously the moderating effect of both consumers' knowledge about the luxury brand and product ephemerality as well as the mediating effect of NGLP-luxury brand fit. Therefore, in this experiment a "full" model considering all factors investigated in this research was employed in comparison to the study by De Angelis et al. (2017). Overall, results of this experiment indicate that design similarity strategy, product ephemerality and luxury brand knowledge jointly influence NGLP purchase intention though fit perceptions. In essence, this experiment suggests that luxury brands should adopt a different design approach when introducing a NGLP, depending on the type of consumer being targeted and the type of product being introduced. Specifically, consumers with low or medium levels of luxury brand knowledge tend to negatively evaluate luxury companies' introduction of a NGLP looking similar in design to models of a green company only when the product is ephemeral due to reduced fit perceptions. In contrast, when the product is durable, design similarity has been shown not to have a significant effect on purchase intention at any level of luxury brand knowledge (low, medium or high).

Theoretically, this study contributes to three different research streams. First, it advances knowledge into the drivers of new green product innovation and acceptance by studying a previously overlooked factor (product design). Specifically, the present research proposes that the benchmark that companies use when designing their new green products (i.e., their previous models or the models of green-oriented companies) may lead to different results fostered by different evaluations of the fit between the new product and the luxury brand introducing it.

Second, this investigation offers a significant contribution to the literature on sustainable consumption, which has overlooked the luxury segment in favour of mass market segments, such as food and cosmetics. As a consequence, most past research suggests that luxury companies are unlikely to succeed in their sustainability initiatives. This motivated the third contribution of this research, which was to propose and show that it is indeed possible to reconcile luxury and sustainability. Specifically, results presented in this article highlight that the design of NGLPs can influence consumers' inclination to embrace them, and that such an effect importantly depends on both consumer- and product-related factors.

From a managerial standpoint, this research presents interesting implications for luxury companies that want to increase the effectiveness of their sustainability initiatives. While luxury brands devote considerably more attention to environmental protection today than in recent decades, such brands still struggle with understanding how to improve—or at least avoid reducing—consumers' receptiveness to their green new models. Our study suggests that product design plays a pivotal role in consumers' attitudes and behaviours toward sustainable luxury. In particular, companies should pay special attention to the aesthetic characteristics of their green models, generally aligning the design with the luxury brand's image rather than with the existing models manufactured by companies that specialize in green production (which might be perceived as lower quality). Additionally, our findings suggest that luxury companies focused on the production of more durable goods, may have greater design freedom than their counterparts involved in ephemeral luxury products. In our study, ephemeral products were perceived more negatively when they resembled a model from a green company rather than one from the luxury company.

In short, it seems that luxury companies should engage in careful research and segmentation analysis before introducing green new models to the market. Specifically, luxury companies should weigh the type of consumer (established clients vs. potential clients, expert vs. inexpert people) and the product category (characterized more by ephemerality vs. durability) when launching a green new product. In essence, with a focus on the luxury segment, this study provides full support to the claim that smarter design decisions can promote sustainable consumption (Küçüksayraç 2015).

Our study features some limitations that may provide avenues for future investigation. First, we focused on one specific dimension along which green new products may vary, namely their similarity in design to established models. Future research could explore the relationship between consumers' purchase intentions and other product dimensions. Second, our research provides novel and converging evidence about the effect of green product introduction strategies on consumers' evaluations and purchase intentions; however, it did not investigate the underlying theoretical mechanism behind this effect. Thus, future work might attempt to isolate the psychological aspects associated with the type of strategy chosen. Third, we focused on two specific moderators—namely, consumers' luxury brand knowledge and level of product ephemerality—but future studies could certainly investigate how other factors might act as moderators of the effect described in this research. Last, while in all our experiments we recruited online respondents, future research could test if our results hold when recruiting actual luxury consumers via field experiments.

Appendix

Stimulus Material Used in the Experiment

Condition #1: Similar to the luxury company's previous models with the durable product

> The brand Armani, a leading luxury brand, presents a new wallet model that is **very long-lasting (it will last for decades)** and classic (it will never go out of fashion). Moreover this new wallet is also eco-friendly. Indeed, this wallet is made by eco-friendly materials that are alternative to regular materials typically used in the production of wallets.
> This wallet is quite **similar to the previous typical models of wallet produced by the same brand** Armani in terms of design (i.e., how it looks like).

Condition #2: Similar to the luxury company's previous models with the ephemeral product

The brand Armani, a leading luxury brand, presents a new haute couture jacket from its latest collection that follows this season's fashion trends, thus being **quite seasonal** and not classic (**it will soon go out of fashion**). Moreover this new jacket is also eco-friendly. Indeed, this jacket is made by organic textiles that have very low impact on the environment.

This jacket is quite **similar to the previous typical jackets of the same brand** Armani in terms of design (i.e., how it looks like).

Condition #3: Similar to the green company's previous models with the durable product

The brand Armani, a leading luxury brand, presents a new wallet model that is very **long-lasting (it will last for decades)** and classic (it will never go out of fashion). Moreover this new wallet is also eco-friendly. Indeed, this wallet is made by eco-friendly materials that are alternative to regular materials typically used in the production of wallets.

This wallet is quite **similar to the wallets produced by the green brand GreenFrames** in terms of design (i.e., how it looks like). GreenFrames is one of the few wallet brands with a specific core focus on environmental sustainability.

Condition #4: Similar to the green company's previous models with the ephemeral product

The brand Armani, a leading luxury brand, presents a new haute couture jacket from its latest collection that follows this season's fashion trends, thus being **quite seasonal** and not classic (**it will soon go out of fashion**). Moreover this new jacket is also eco-friendly. Indeed, this jacket is made by organic textiles that have very low impact on the environment.

This jacket is quite **similar to the jackets produced by the green fashion brand GreenFashion** in terms of design (i.e., how it looks like). GreenFashion is one of the few fashion brands with a specific focus on environmental sustainability.

References

Achabou MA, Dekhili S (2013) Luxury and sustainable development: is there a match? J Bus Res 66(10):1896–1903

AFP (2008) WWF épingle l'industrie du luxe sur l'environnement. Agence France Presse

Albino V, Balice A, Dangelico RM (2009) Environmental strategies and green product development: an overview on sustainability-driven companies. Bus Strateg Environ 18(2):83–96

Atwal G, Williams A (2009) Luxury brand marketing–the experience is everything! J Brand Manage 16(5):338–346

Bain and Company (2014) Luxury goods worldwide market study. http://www.bain.com/bainweb/PDFs/Bain_Worldwide_Luxury_Goods_Report_2014.pdf. Accessed 7 Mar 2017

Beckham D, Voyer BG (2014) Can sustainability be luxurious? A mixed-method investigation of implicit and explicit attitudes towards sustainable luxury consumption. Adv Consum Res 42:245–250

Berthon P, Pitt L, Parent M et al (2009) Aesthetics and ephemerality: observing and preserving the luxury brand. Calif Manage Rev 55(1):45–66

Boenigk S, Schuchardt V (2013) Cause-related marketing campaigns with luxury firms: an experimental study of campaign characteristics, attitudes, and donations. Int J Nonprofit Voluntary Secur Mark 18(2):101–121

Cho YN, Thyroff A, Rapert MI et al (2013) To be or not to be green: exploring individualism and collectivism as antecedents of environmental behavior. J Bus Res 66(8):1052–1059

Chung Y, Tsai C (2007) The effect of green design activities on new product strategies and performance: an empirical study among high-tech companies. Int J Manag 24(2):276–288

Creusen MEH, Schoormans JPL (2005) The different roles of product appearance in consumer choice. J Prod Innovat Manag 22(1):63–81

Cucuzzella C (2016) Creativity, sustainable design and risk management. J Clean Prod 135(1):1548–1558

Dangelico RM, Pontrandolfo P (2010) From green product definitions and classifications to the green option matrix. J Clean Prod 18(16):1608–1628

Davies IA, Lee Z, Ahonkhai I (2012) Do consumers care about ethical-luxury? J Bus Ethics 106(1):37–51

De Angelis M, Adıgüzel F, Amatulli C (2017) The role of design similarity in consumers' evaluation of new green products: an investigation of luxury fashion brands. J Clean Prod 141:1515–1527

Dodds WB, Monroe K, Grewal D (1991) Effect of price, brand and store information on buyers' product evaluations. J Mark Res 28(3):307–319

Du S, Bhattacharya CB, Sen S (2007) Reaping relational rewards from corporate social responsibility: The role of competitive positioning. Int J Res Mark 24(3):224–241

Dubois B, Laurent G (1998, June) The new age of luxury living. Finan Times Mastering Manag Rev 32–35

Esch FR, Langer T, Schmitt BH et al (2006) Are brands forever? How brand knowledge and relationships affect current and future purchases. J Prod Brand Manage 15(20):98–105

Fletcher K (2013) Sustainable fashion and textiles: design journeys. Routledge, London

Fuchs C, Prandelli E, Schreier M et al (2013) All that is users might not be gold: How labeling products as user designed backfires in the context of luxury fashion brands. J Mark 77(5):75–91

Gershoff AD, Frels JK (2015) What makes it green? The role of centrality of green attributes in evaluations of the greenness of products. J Mark 79(1):97–110

Griskevicius V, Van den Bergh B, Tybur JM (2010) Going green to be seen: status, reputation, and conspicuous conservation. J Pers Soc Psychol 98(3):392–404

Gürhan-Canli Z, Maheswaran D (1998) The effects of extensions on brand name dilution and enhancement. J Mark Res 11(1):464–473

Han YJ, Nunes JC, Drèze X (2013) Signaling status with luxury goods: The role of brand prominence. Int Retail Mark Rev 9(1):1–22

Haws KL, Winterich KP, Naylor RW (2014) Seeing the world through GREEN-tinted glasses: green consumption values and responses to environmentally friendly products. J Consum Psychol 24(3):336–354

Hayes AF (2013) Introduction to mediation, moderation, and conditional process analysis. The Guilford Press, New York

Hayes AF, Preacher KJ (2014) Statistical mediation analysis with a multicategorical independent variable. Brit J Math Stat Psy 67:451–470

Janssen C, Vanhamme J, Lindgreen A et al (2014) The catch-22 of responsible luxury: Effects of luxury product characteristics on consumers' perceptions of fit with corporate social responsibility. J Bus Ethics 119(1):45–57

Janssen MA, Jager W (2002) Stimulating diffusion of green products: co-evolution between firms and consumers. J Evol Econ 12:283–306

Johnston J, Szabo M, Rodney A (2011) Good food, good people: understanding the cultural repertoire of ethical eating. J Consum Cult 11(3):293–318

Kapferer JN (1998) Why are we seduced by luxury brands? J Brand Manag 6(1):44–49

Kapferer JN (2010) All that glitters is not green: the challenge of sustainable luxury. Eur Bus Rev, 40–45. Retrieved from http://www.theluxurystrategy.com/site/wp-content/uploads/2011/01/EBR.JNK_.NovDec2010_All-that-Glitters-is-not-Green.SustainableLuxury.pdf. Accessed 7 Mar 2017

Kapferer JN, Michaut-Denizeau A (2014) Is luxury compatible with sustainability? Luxury consumers' viewpoint. J Brand Manag 21(1):1–22

Kirmani A, Sood S, Bridges S (1999) The ownership effect in consumer responses to brand line stretches. J Mark 63(1):88–101

Kronrod A, Grinstein A, Wathieu L (2012) Go green! Should environmental messages be so assertive? J Mark 76(1):95–102

Küçüksayraç E (2015) Design for sustainability in companies: strategies, drivers and needs of Turkey's best performing businesses. J Clean Prod 106(1):455–465

Lin YC, Chang CA (2012) Double standard: The role of environmental consciousness in green product usage. J Mark 76(5):125–134

Ljungberg LY (2007) Materials selection and design for development of sustainable products. Mater Des 28(2):466–479

Lloret A (2016) Modeling corporate sustainability strategy. J Bus Res 69(2):418–425

Lochard C, Murat A (2011) Luxe et développement durable. Eyrolles-Éd. d'Organisation, Paris

Loken B, John DR (1993) Diluting brand beliefs: when do have a negative impact? J Mark 57 (3):71–84

Lozano R, Huisingh D (2011) Inter-linking issues and dimensions in sustainability reporting. J Clean Prod 12:99–107

Magnoni F, Roux E (2012) The impact of step-down line extension on consumer-brand relationships: a risky strategy for luxury brands. J Brand Manag 19(7):595–608

Martinez E, Polo Y, De Chernatony L (2008) Effect of brand extension strategies on brand image: A comparative study of the UK and Spanish markets. Int Mark Rev 25(1):107–137

McKinsey and Company (2008) Helping 'green' products grow. http://www.data360.org/pdf/20081029174901.08-10-29%20McKinley%20Green%20Perception.pdf. Accessed 7 Mar 2017

McKinsey and Company (2011) The business of sustainability: McKinsey global survey results. http://www.mckinsey.com/insights/energy_resources_materials/the_business_of_sustainability_mckinsey_global_survey_results. Accessed 7 Mar 2017

Milberg SJ, Park CW, McCarthy MS (1997) Managing negative feedback effects associated with brand extensions: the impact of alternative branding strategies. J Consum Psychol 6(2):119–140

Ngobo PV (2011) What drives household choice of organic products in grocery stores? J Retail 87 (1):90–100

Nia A, Zaichkowsky JL (2000) Do counterfeits devalue the ownership of luxury brands? J Prod Brand Manag 9(7):485–497

OCSC (2000) Report on the second session of the Oxford commission on sustainable consumption. OCSC 2.8, Mansfield College, Oxford Centre for the Environment, Oxford

Olsen MC, Slotegraaf RJ, Chandukala SR (2014) Green claims and message frames: how green new products change brand attitude. J Mark 78(5):119–137

Peattie K, Peattie S (2009) Social marketing: a pathway to consumption reduction? J Bus Res 62(2):260–268

Pickett-Baker J, Ozaki R (2008) Pro-environmental products: marketing influence on consumer purchase decision. J Consum Mark 25(5):281–293

Porter ME, van der Linde C (1995) Green and competitive: ending the stalemate. Harvard Bus Rev 73(5):120–133

Preacher KJ, Rucker DD, Hayes AF (2007) Addressing moderated mediation hypotheses: Theory, Methods, and Prescriptions. Multivariate Behav Res 42(1): 185–227

Pujari D (2006) Eco-innovation and new product development: influences on market performance. Technovation 26:76–85

Schlegelmilch BB, Bohlen GM, Diamantopoulos A (1996) The link between green purchasing decisions and measures of environmental consciousness. Eur J Mark 30(5):35–55

Sen S, Bhattacharya CB (2001) Does doing good always lead to doing better? Consumer reactions to corporate social responsibility. J Mark Res 38(2):225–243

Shrum LJ, McCarty JA, Lowrey TM (1995) Buyer characteristics of the green consumer and their implications for advertising strategy. J Advertising 24(2):71–82

Smith RA (2009, January) The $43,000 recession suit: even as overall sales wane, some luxury makers push the price envelope: 'On really special thing'. Wall Street J 24

Steinhart Y, Ayalon O, Puterman H (2013) The effect of an environmental claim on consumers' perceptions about luxury and utilitarian products. J Clean Prod 53:277–286

Stock A, Balachander S (2005) The making of a 'hot product': a signaling explanation of marketer's scarcity strategy. Manage Sci 51(8):1181–1192

Strong C (1997) The problems of translating fair trade principles into consumer purchase behaviour. Mark Intell Plann 15(1):32–37

Sujan M, Dekleva CA (1987) Product categorization and inference making: some implications for comparative advertising. J Consum Res 14(3):372–378

The Boston Consulting Group (2015) True luxury global luxury consumer insight. http://www.bcg.it/documents/file181201.pdf. Accessed 7 Mar 2017

Torelli CJ, Basu-Monga S, Kaikati A (2012) Doing poorly by doing good: corporate social responsibility and brand concepts. J Consum Res 38(5):948–963

Vandermerwe S, Oliff M (1990) Customers drive corporations green. Long Range Plan 23(6):10–16

Van Doom J, Verhoef PC (2011) Willingness to pay for organic products: differences between virtue and vice foods. Int J Res Mark 28(3):167–180

Vigneron F, Johnson LW (2004) Measuring perceptions of brand luxury. J Brand Manage 11:484–506

Ward D, Chiari C (2008) Keeping luxury inaccessible. MPRA Paper No. 11373, posted 20 November 2008, Accessed 7 Mar 2017, University Library of Munich, Germany

WCED (1987) Our Common Future (The Brundtland Report), World Commission on Environment and Development. Oxford University Press, Oxford

Widloecher P (2010) Luxe et développement durable: Je t'aime, moi non plus? Luxefrancais

Worcester R (1993) Public and elite attitudes to environmental issues. MORI, London

Yang D, Lua Y, Zhub W et al (2015) Going green: how different advertising appeals impact green consumption behaviour. J Bus Res 68(12):2663–2675

The Carloway Mill Harris Tweed: Tradition-Based Innovation for a Sustainable Future

Thomaï Serdari

Abstract Textile manufacturing, and within it Harris Tweed production, is as deeply ensconced into Scottish heritage as champagne is to French culture. Today, there are three Mills in Scotland that produce Harris Tweed (twill) fabric, which is then marketed worldwide for production of luxury clothing. The Harris Tweed industry, initially based exclusively on hand-made processes (dyeing, spinning, and weaving) was transformed in the mid-19th century into a more standardized process, still hand-crafted, and as impeccable as machine-made production methods. This led to the booming of demand for Harris Tweed, which essentially made that industry the base of the Outer Hebrides of Scotland's economy through the 1970s. The decline of European textile industries that followed (Harris Tweed included) and the economic depression that ensued only proves how inextricably linked the local Scottish economy is to this unique type of craftsmanship. The Carloway Mill, one of the three remaining Mills, is under new leadership since 2005. The last ten years have proven critical for the Mill. Harris Tweed survived primarily because of its distinct traits and quality but also "because it is protected by an Act of Parliament limiting the use of the Orb trademark to hand woven tweeds made in the Outer Hebrides of Scotland." In addition to governmental intervention, the new leadership took a bold approach to solving the issue of declining demand for the luxurious fabric by catalyzing an internal product development process. In other words, an industry that has relied on the same type of fabric for the last two hundred years has been revitalized from within and with the launch of the new lighter version of Harris Tweed invented exclusively at Carloway Mill. It is at that Mill that the art of weaving Harris Tweed, the culture of the local craftsmen, and the innovation that comes with a 21st century perspective on textile functionality will all contribute to the legacy of the evolving local craftsmanship. The impact of this innovative product on the local community and its economy is of tremendous significance as it also marks a transformative period in which sustainability is what drives the locals' survival: they either introduce more positive changes into a legacy

T. Serdari (✉)
Leonard N. Stern School of Business, New York University, Tisch Hall,
40 West Fourth Street, 800, New York, NY 10012, USA
e-mail: tserdari@stern.nyu.edu

© Springer Nature Singapore Pte Ltd. 2018
M. A. Gardetti and S. S. Muthu (eds.), *Sustainable Luxury, Entrepreneurship,
and Innovation*, Environmental Footprints and Eco-design of Products
and Processes, https://doi.org/10.1007/978-981-10-6716-7_10

process of textile production or they suffer the financial and social repercussions that stagnation entails.

Keywords Craftsmanship · Entrepreneurship (Responsible) · Sustainability Luxury

1   Introduction

The Harris Tweed industry, for most considered a remnant of the past, presents an opportunity to study its elements anew and reframe them both within the context of luxury and that of sustainability. This 700 year-old manufacturing sector continues to adjust and evolve while preserving its traditional elements. These link it to its land of origin, its people, and to the human spirit. The industry's structure is germane to the discussion of slow fashion and, beyond that, slow culture—the only way to a sustainable future.

On the contrary, textile innovation today is primarily linked to synthetic materials, woven to produce high-performance fabrics that may even consist of microchips to gather data from the wearer's body. Moving at a galloping pace, this fascinating, technology-based, large-scale industrial sector of high pollutants caters to the individual at the expense of the collective.

At the Carloway Mill, one of the three remaining "homes" where Harris Tweed is handcrafted, we will study the strengths, weaknesses, threats and opportunities that have pushed a small enterprise to experiment with tradition in order to ensure a sustainable future for itself and its people. Specifically, we will try to understand how the Harris Tweed industry arrived to where it is today and what it means for the Carloway Mill to have introduced a "new" type of fabric that addresses the modern wearer's concerns about climate change, environmental pollution, and zero-waste fashion design.

In addition to various scholarly articles and books that were consulted, business intelligence was gathered by commercial and popular press perusal, including commercial brands' websites. *The New Yorker* and *Vogue* archives offered testimony to the market's swings and consumers' tastes through the decades. Both databases cover the same time period, from the end of the nineteenth century to today. Two types of material were isolated and examined: paid advertisements (for size, placement, word count, image choice, and creative approach) and editorials in which Harris Tweed is featured (for frequency and constancy). Finally, this business case would not have been written without a series of interviews that I conducted over a period of almost six months, from October 2016 through March 2017, with Derek Reid, former CEO of the Carloway Mill, Alan Bain, Director, Annie Macdonald, former Head of Operations and current CEO of Carloway, and "Biddy" (Murdo McLeod), Head Hand-warper and Creative Director at Carloway. I am most grateful to all four for their time, patience, and astute insights about their own processes and about the textile industry as a whole on a global scale.

The ensuing discussion will serve as the point of departure for several marketing projects, the difficulty of which rests on the fact that we cannot afford not to produce luxury products. They present the only viable solution to applied sustainability. This depends on enterprises that place emphasis on: remaining harmless as opposed to reversing or counterbalancing their harmful operations; empowering people as opposed to forcefully intervening with their future; and trying to learn from nature as opposed to trying to tame her.

2 Place: The Outer Hebrides

The long and narrow body of water, known as the Minch, that separates the Outer Hebrides from the Scottish mainland is a portal to a place where time stands still. About 200,000 visitors choose to cross that invisible divide every year, mostly on vacation, by air and predominantly by ferry (Snedden Economics 2007). On the other side, the many islands of the Outer Hebrides await them. They form an archipelago of five main islands: Lewis and Harris, North Uist, Benbecula, South Uist, and Barra and several uninhabited ones, all of them perforated by large bodies of water, known as loch (Murray 1973). The landscape appears harsh and wind beaten. The hilly land rises in unexpected contortions as if resisting the brutal winds only to descend in fertile low-lying dunes. The exposed rock of the islands hovers over a sandy coastline. The landscape appears rugged and unpredictable, much like the wind. Yet the temperatures remain temperate throughout the year, greatly benefiting from the North Atlantic Current (Haswell-Smith 2004). Both land and people seem caught in between the sweetness of the climate and the violence of the winds. The area's proximity to the North Pole ensures long days during the summer, which is usually dry and lasts from May to August. Against the nakedness of the land, the Callanish Stones on the island of Lewis have been standing in a cruciform pattern with an interior stone circle since about 2900 BC, also known as the Bronze Age, and greet the visitors without other references to time. In the Outer Hebrides time does not exist. Only land formations of rock, sand, and grassy plains (*machair*), the water, and the wind exist. The archipelago caught in a permanent state of turbulence does not know time.

The residents, who had flocked to the islands in older times to escape extreme circumstances, have been steadily abandoning their land, seeking opportunities and tamer forms of life in mainland Scotland or elsewhere in the United Kingdom (Thompson 1969). Those who remain are the offspring of four main clans, (MacLeods, MacDonalds, MacKenzies, and MacNeils) and speak both Scottish Gaelic and modern English but prefer the former (Buxton 1995). Fishing which used to be the main form of employment has been steadily declining as an industry. Crofting, a local form of land tenure for small-scale food production, is still prevalent and provides both for humans and their livestock for which poorer quality *machairs* are communally owned and used for grazing (Byron and Hutson 1999). The textile industry is the only other option, aside from tourism, that can provide steady employment to the Outer Hebrides' people. They have been producing woolen cloth for many centuries and continue their craft today.

3   Heritage: The Cloth Industry

Weaving is one of the most ancient industries. While cloth does not survive easily, archaeologists have identified cloth that dates to the Bronze Age, if not earlier. Such is the case in Scotland as well where remnants of ancient cloth have been unearthed along with combs. These combs were used for weaving in the manufacture of fabrics and have occasionally been found with spinning whorls that aid in spinning the wool into yarn. According to *Senchus Mor*, a 1000 year old manuscript that historian Francis Thompson consulted for his account of the Harris Tweed industry, cloth-making has involved spindles, spinning-stick, wool-bag, weaver's reed, distaff-spool stick, flyers, needles, beams, swords or weaving sticks and *glaisin* dye. The ancient manuscript refers to cloth making as a highly skilled craft that requires artistry (Thompson 1969).

Mountain sheep, the Scottish Blackface and the Long-faced or White-faced sheep, known today as the Cheviot sheep, were originally introduced into Scotland around 1372 (Thompson 1969). The two breeds have provided wool for Scottish cloth through the centuries. The Blackface fleece tends to be coarser and longer with many more impurities while the Cheviot fleece is shorter, finer, and, as a result, cleaner. Wool production begins with washing in scouring water and soaps several times over and until any fats and impurities are completely removed. Rinsed with cool water the fleece is laid out on grass and out of direct sunlight to dry. Dyeing is done while the wool is still in a mass and before carding.

Dyeing is a craft as ancient as weaving and requires both scientific knowledge and artistry. In Scotland, and particularly the Outer Hebrides, dyes originate from local flora and natural ingredients that preserve the coloration of the surrounding landscape. The deep knowledge about indigenous plants' properties has been passed down from generation to generation, usually from mother to daughter who contribute in the manufacturing of the tweed woolen cloth, or as it was known in Scotland, *clo mor* (Thompson 1969). For example, blue can be made from elderberry mixed with alum; yellow from buckthorn or bracken root; red from stone parmelia, a flat plant with a black underside; black or grey from water-flag root or yellow flag iris; dark green from iris leaf, broom, and whin bark; magenta from dandelion; orange from barberry root or peat soot; purple from cudbear; black from the boiling of the briar bark, oak or alder. This list is not exhaustive and is indicative of how much of the character of the tweed depends on the land, the weather conditions, and the particularities of each island's microenvironment.

Carding, namely a careful scarping between two flat hard boards covered with strong, wired teeth embedded in leather, follows the dyeing process. This is a major difference of modern day industrial production of yarn where dyeing occurs only after the yarn has been created. Carding loosens up the wool and softens it, ultimately renders it thinner and thinner so that the spinning can begin. With spinning the wool turns into yarn. Whorl, spindle and wheel are instruments with which we are well acquainted even today as they survived well in the nineteenth century. Once the yarn has been formed it is arranged in the proper order by color (warping) so that it produces the desired pattern in the web. Warping was traditionally done by women (Thompson 1969). Weaving tweed was an arduous and complex process. It involved using both feet and hands and required focus and dexterity. Depending on the operator's efficiency, the quality and quantity of the yarn, the complexity of the pattern of the cloth, the production of a web of tweed may have required more than a week of continuous effort. Worn by unassuming locals (fishermen and shepherds) tweed production was intricate and expensive. The weave attained at the end of that week was not even ready for use. It required *waulking*, a combination of five sequential stages, thickening of the cloth, cleansing, folding, giving tension to the cloth, and the rite of consecration of the cloth. The women who conducted these processes were singing special songs to mark the ceremonial character of their work. Several accounts of traditional waulking have been recorded by prominent visitors of the Hebrides such as Dr. Samuel Johnson or Sir Walter Scott and they survive in their biographies (Thompson 1969). The quality of the final product was marked as exceptional. Ironically, while the diet of the Hebrideans was relatively limited, they were all fairly well dressed.

In the beginning of the nineteenth century several aspects of the weaving process evolved to a better type of warping which continued to require nevertheless, as it does today, an advanced set of motor skills from the operator (Anderson 2013). The evolution allowed for the home industry to grow and find new markets on the mainland (Scotland and England). As the demand for tweed grew, more women were absorbed in the industry while their husbands had turned to the quite lucrative industry of herring fishing. Tweed producers were supported by non-profit groups such as the Scottish Home Industries Association and encouraged to establish sales contacts mostly through visitors who were moving regularly between the Mainland and the Isles. In 1849, Harriet, Duchess of Sutherland established an Industrial Society which later became Sutherland Home Industries. In 1889, the Scottish Home Industries Association was established under the patronage of Princess Louise (Thompson 1969). It is no coincidence that during that time, the Arts and Crafts Movement was gaining momentum as a reaction to the sweeping changes that came with the industrial revolution earlier in the same century. People were already regretting the dominance of the machine as a means of production, the uncontrollable growth of cities, and the alienation from land and home, especially pronounced amongst industry workers. The march toward overproduction and overconsumption had already begun. The tweed industry seemed like a haven of sustainable processes that respected both environment and man. In addition, they were exemplary of local traditional modes of making and maintained a unique artistic character that echoed the flora and fauna of the Outer Hebrides (Moisley 1961).

The end of the nineteenth century brought a major change in the tweed industry. Carding that had been done by hand, rendering production slow and extremely tiring, changed to a machine-aided process. Several producers began sending their cloth to the mainland to be carded at mills there. This however was not ideal as it took production away from its origin and made it confusing for customers to understand whether their favorite woolen cloth was indeed manufactured by hand or by machine (Anderson 2000). As a response, the first carding mill was erected in Harris in 1900 by Sir Samule E. Scott, the proprietor of the North Harris estate (Thompson 1969). Three years later, a new mill was erected in Stornoway. In 1904, Diricleit in Harris, the third mill, was built. Inspite of the expediency of production facilitated by the mills, islanders were not able to meet the growing demand for tweed cloth, a fabric favored by royalty and other upper class citizens. The demand for tweed had become so great that in 1906 Henry Lyons of London was convicted of fraud, namely "selling a Harris Tweed suit when in fact the cloth was mill-spun and power-loomed in Huddersfield" (Thompson 1969). This was an issue of economics and not merely pride (Herman 1957). In 1911, a tweed cloth made by machine-spun yarn made profit of about £3 per web whereas authentic tweed (hand-made) made just a fraction of that profit because of the increased cost of production. Discussions led to the organization of the Outer Hebrides producers and marketers of authentic Harris Tweed. This first association had a main priority: the application of standardization marks (what is known today as certification trade marks) of the only cloth legitimately described and sold as Harris Tweed.

4   Authenticity: The Harris Tweed Authority

"Woven with patience, love and care" Harris Tweed Authority

What is known today as the Harris Tweed Authority (HTA) was founded on December 9, 1909 as the Harris Tweed Association. Its mission was defined as:

> …the protection of the interests of manufacturers and merchants of and dealers in tweed made in the Islands of Harris, Lewis and Uist in Scotland, and to promote the manufacture and sale of such tweed. To protect the trade against offences under the Merchandise Marks Acts and otherwise to prevent the use of false trade marks and descriptions in respect of tweed made in imitation thereof. (Thompson 1969)

Additionally, the company would have complete control over the trademark of Harris Tweed, market the merchandise in the media, and legally defend manufacturers and merchants against any entity imitating the Harris Tweed product.

As the industry grew so did the Association that came to be known the Association for the Protection of the Harris Tweed Industry. It promoted and protected the mark of Harris Tweed so as for it to mean "hand spun, hand woven, dyed and finished by hand on the Islands of Lewis, Harris, Uist, Barra and their several purtenances and all known as the Outer Hebrides" (Thompson 1969). The Association began stamping tweed with its trademark in 1911. The production capacity of each island varied but detailed records were kept under the Association's watch.

In the 1930s, after the industry had grown considerably, the Association failed to perform the duties promised to its members, especially in terms of promotion. Additional disagreements rose between the Association and its members because yarn was widely mill spun as opposed to hand spun, a fact that the Association resisted to recognize. The complexity of the issue led to market confusion, a depression that was felt by all and a substantial price drop for the hand-woven tweed. It took three additional years to resolve the disagreement which concluded with the inclusion of "hand-woven" in the definition of the Harris Tweed name and the introduction of the new term "at their own homes" which would ensure that the hand-weaving would remain a cottage industry. The recovery took full effect in two years and by 1935 the islands had six mills (for spinning, carding, dyeing, and finishing) four in Stornoway and two in Harris. While the Second World War slowed things down, the industry recovered in the 1950s and the Association continued the promotion of the product in foreign markets as well, the US included. It is recorded that in the 1960s some 7.6 million yards of Harris Tweed were produced and stamped as opposed to the few hundreds at the beginning of the century.

It was the early 1990s when another major transformation of the industry took place based mainly on the increase of orders from luxury houses that were discovering the traditional fabric and making it fashionable again in new and unexpected ways. Production adjusted to a new double width look which actually required retraining the weavers. This was a great opportunity for the Harris Tweed Association to impose stricter quality standards. By an Act of Parliament in 1993, the Harris Tweed Authority took over from the Harris Tweed Association and the brand of Harris Tweed became statutory forever linked to the Hebrides Islands. "The mark of the Orb, pressed onto every length of cloth and seen on the traditional label affixed to finished items, guarantees the highest quality tweed, dyed, spun and hand-woven by islanders of the Outer Hebrides of Scotland in their homes to the laws outlined in the Harris Tweed Act of Parliament" (https://www.harristweed.org/about-us/guardians-of-the-orb/).

The Harris Tweed Authority oversees the production that takes place in the three remaining mills, Harris Tweed Hebrides, Kenneth MacKenzie, and Carloway Mill as well as the seventeen independent producers of Harris Tweed who also produce cloth for the mills.

5 Enterprise: The Carloway Mill and Its Product

Craftsmanship for today and tomorrow's world

One of the three surviving mills, the Carloway Mill was revived with its purchase by three investors, Derek Reid, Alan Bain, and Roddy MacAskill in 2003 (Figs. 1 and 2). While all three are Scots, MacAskill had been a local Harris businessman and Bain a New York lawyer. Having had a long and successful career as a CEO of a major company in the consumer goods industry, Reid assumed the role of overseeing manager at Carloway. It was not the mill's financial health that

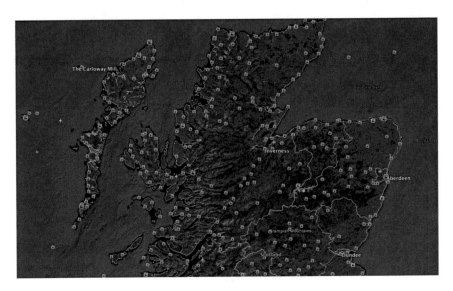

Fig. 1 The outer Hebrides. Carloway. *Courtesy* Google Maps

Fig. 2 The Carloway Mill. Street View. *Courtesy* Google Maps

drove the three investors to its acquisition but rather the iconic identity of Harris Tweed and its place in the local culture. Reid had become acquainted with the product and culture of the Western Isles when he served as the CEO of The Scottish Tourist Board between 1994 and 1996. In the years that followed, the industry experienced a decline that propelled the investors to give it a financial boost and substantial personal time and effort.

The first few years at Carloway were focused on reviving the mill with the purchase of new plant equipment, overseeing its installation, and ensuring its compliance with all the standards of the Harris Tweed Authority (Figs. 3, 4, 5 and 6).

From 2003 to 2016 the company grew from 5 to 25 employees while its turnover increased from merely insignificant to £1.5 million. In 2016, Reid assumed the role of CEO and appointed a new manager of operations Annie Macdonald. As the CEO, Reid articulated the following vision for the company:

> We believe that respect for traditional Scottish traditions, perseverance against industry threats, and an open mind will lead us to being a luxury brand within the brand of Harris Tweed.
>
> We believe Carloway Harris Tweed to be synonymous with innovation, superior quality, and adherence to traditional methods of tweed production.
>
> We believe in treating our staff as family and in empowering our weavers in their journey of preserving their craft and transmitting it to the next generation while catalyzing financial prosperity for our region.

Fig. 3 Blended Wool. The Carloway Mill. *Courtesy* Annie Macdonald

Fig. 4 Carding at the Carloway Mill. *Courtesy* Annie Macdonald

Fig. 5 Preparing to be spun at the Carloway Mill. *Courtesy* Annie Macdonald

Fig. 6 Making Tweed at the Carloway Mill. *Courtesy* Annie Macdonald

We believe in the human element of our work and find pride in our product.

Carloway is the smallest of the three mills and an independent wholesale producer of Harris Tweed. Exemplary in its old-style production of the cloth, Carloway owns traditional craft equipment and employs a crew of local home-based weavers for the production of a distinctively different material that befits bespoke orders (Kitchener 1994). The mill weavers are self-employed. When they produce cloth for the mill they are supplied with beamed warps and yarn as well as instructions on the particular pattern to be woven. When the tweed is ready, the mill employee collects it to finish it and stamp it before it is shipped to the customer. According to Macdonald (who is currently Carloway's CEO after a Management Buyout was completed in February 2017) the minimum order is usually 20 m single width (75 cm wide and woven on a Hattersley loom) or 50 m double width (150 cm and woven on a Bonas-Griffith rapier loom). The delivery time depends on the size of the order and can take anywhere between 6 and 12 weeks since each product is made from scratch. Single width has a faster production cycle, 2–3 weeks shorter. Single or double width orders may be delayed if the colors are not readily available. The making of Harris Tweed is based on dyed fleece and not dyed yarn. Therefore, bespoke colors require the mixing of fleece in the basic colors at varied combinations of carefully weighed wool in exact proportions as defined by the recipe for each color (Fig. 7). Dyeing the fleece is a long process in itself and may therefore add to the overall turnaround time for the particular order (Lawson 2011).

Fig. 7 Dalmore Herringbone Tweed. *Courtesy* Annie Macdonalnd, The Carloway Mill

The employees have long experience in the industry having previously worked at Shawbost and Stornoway (as tweed producers). One amongst them, "Biddy" (Murdo McLeod) is one of the three remaining hand-warpers in the industry. Very knowledgeable and talented in hand-warping, "Biddy" is in charge of creative decisions as hand warping is the process that determines the transferability of the individual pattern that has been designed for the particular piece of cloth. Already in his seventies, "Biddy" is eager to transmit his knowledge of "know-how" to a younger generation of weavers, which has proven difficult because hand-warping resembles music skills. One may know how to play the piano but that does not mean that she is a good pianist. At the moment, another one of the Carloway employees is able to divide the workload of hand warping with "Biddy." Four hands are better than two.

Additionally, the management took care to increase the electricity capacity of the mill so that all carding machines can be operative. The production process remains consistent with what has been discussed in the historic account of the industry. It consists of washing of the virgin wool, dying, blending, carding, spinning and beaming. The beams with the yarn are transported to the homes of the weavers. The average time for the yarn to be converted to tweed is 5 days. In total, Carloway can produce up to 50 tweeds or about 4000 m of tweed per week.

When the mill picks up the tweed from the weavers, payment to them must be made (as they are independent contractors). At the mill, the cloth is washed, darned (for quality control) and stamped by the Harris Tweed Authority (HTA) with the official "Orb" that classifies it as genuine. Finally, it is sent to the customer. The process is intense, time consuming, and cash flow heavy. The mill pays for the wool purchased, the dyes, the overheads, the weavers' and administrative personnel's wages, and the HTA. It takes 30 days to recover the cash from the customer. In other words, customers are used to paying only after they receive their tweed, a practice that seems impertinent for a heritage-driven industry.

While Harris Tweed production has been perfected in terms of craftsmanship, the cloth itself still maintains its rough character, which is, generally speaking, more appropriate for men's fashion. This was particularly true in the industry's early days when the heaviness of the fabric combined with its muted colors seemed well matched with men's clothing. Slightly more vibrant colors appealing to women were introduced after the Second World War and developed through the 1950s. Today, coloration is not an issue as all mills can be very creative in the ways they mix the colored fleece to achieve the desired hue. However, fabric coarseness and heaviness is an issue in the sense that it limits the uses of tweed within the fashion industry. Appropriate for outwear and even shoes, as proven by Nike's launch of the "AirRoyal Harris Tweed" model (Fig. 8), tweed has had limited applications and no change in the last 30 years.

This lack of innovation motivated Carloway's management to delve deeper into the area of product development. Their work revolved around the desirability of the product and resulted in the softest and lightest tweed produced until now. They researched various combinations of types of fleece and concluded in one from Australian sheep. The new product presents an opportunity to enter women's fashion more aggressively and at the same time differentiates Carloway as an enterprise from the other two mills. Additionally, as Reid remarked during one of our conversations, his team took into account that climate change is progressing rapidly, rendering our climates warmer and the weather patterns unpredictable.

Fig. 8 Nike AirRoyal. Website Screenshot High Snobiety (http://www. highsnobiety.com/)

Such is the environmental reality of our time that it seems absolutely necessary to rethink Harris Tweed for a new world and a customer with different expectations.

6   Threats from Industry Changes

The cyclicality of the fashion industry is a well-known factor that greatly impacts the bottom line of even the most robust companies. Based on the most refined intelligence on future trends and customer preferences, the swings of the market can be anticipated only up to a certain degree. Harris Tweed has been a classic staple in the repertory of dressmakers but its popularity has had its ebbs and flows. To observe this phenomenon archival material was consulted in two American publications: The New Yorker and VOGUE. The former is an American culture magazine in which the HTA placed advertisements for the promotion of the tweed with the American public. The latter is an international fashion magazine that dictates proper fashion and sets trends in large scale. Founded in the US in 1892, the American publication is the earliest and most venerated voice in fashion circles. The British version was launched in 1916 pioneering the fashion magazine's international expansion. Therefore, we consider these to be great primary sources of information in terms of editorial preferences and perceived public influence through paid advertisements. It would be a mistake not to account for personal taste even though this is largely shaped by the trade publications in which journalists inform and instruct the public what to wear. Personal taste relates to a person's: A. *Subjective* experience and aspiration to resemble a personal fashion ideal; B. *Objective* understanding of the functional characteristics of the garment (i.e. is it suitable for sports? Or is it made for a leisurely walk in winter? etc.); and C. *Symbolic* value vis à vis the segment of society in which the person belongs (Berthon et al. 2009).

The aforementioned archives confirmed that Harris Tweed was popular and regularly advertised between 1893 and 1900 but hardly mentioned between 1900 and 1909. This coincides with the reorganization of the industry and the founding of the Harris Tweed Association. The brief drop of interest was replaced by mentions of and print advertisements about the qualities of Harris Tweed in the following three decades. Interest peaked between 1950 and 1959 only to steadily drop until 1970. During the 1970s the print ads became smaller but the number of editorials on Harris Tweed soared. A dramatic drop followed in the 1980s and 1990s. Awareness about the industry increased in the early 2000s, when Chanel openly publicized its relationship with the Scottish mills. There could not have been a more dramatic decline in interest however than what the industry has experienced from about 2010 forward. This last decade is characterized by lack of vigilance as far as the HTA is concerned, a catastrophic reduction of wholesale prices, and a spike in development and production of new synthetic performance fabrics.

In the current highly competitive environment of fast fashion, suppliers have lost their power. Globalization has allowed many new players to enter the market and has unevenly shifted all the bargaining power to buyers (brand manufacturers) who have direct access to customers. Wholesalers are doomed to fail. This is the present state of the market for tweed. Still competing in wholesale, the three remaining tweed mills are mainly competing against new types of textiles, usually machine made and not necessarily natural, that represent just a small fraction of the final garment's cost. In a perpetual cycle, the fashion engine is fed cheap (and often harmful to the planet) raw material to churn out inexpensive and expendable items that have a life cycle of about four weeks. The shorter these cycles become, the greater consumers' appetite for more new items of low quality and short closet life span. Additionally, cheap garments are overproduced and end up in landfills making the issue of fashion waste one of the most threatening of our times (Black 2010).

Street style and athleisure (=athletic + leisure) are two new fashion categories that stem from a strong cultural shift that rejects twentieth century traditional clothing in favor of more relaxed, youthful and active lifestyles. The former is anchored in grassroots urban wear and was initially spotted within the hip hop and skaters' communities but has in the last twenty years gained substantial following. Urban fashion built of mainly heavy cotton items in the form of sweatpants, hoodies, t-shirts etc. has by now become mainstream, greatly impacting consumers' stylistic choices. In winter 2017, one of the most prominent urban fashion labels, *Supreme*, collaborated with luxury fashion house *Louis Vuitton* for the launch of a special collection (Chen 2017). In other words, street style has penetrated fashion tastes in great depth.

Additionally, the emphasis on athletic lifestyles, performance wear, and new forms of exercising has created "athleisure," a new category of clothing based on the concept of comfort, durability, and flexibility. The push generated by the growth of the fitness industry in the US and abroad, the intensification of international travel across time zones, and the slow break down of the formerly austere and constricting corporate structure has fortified consumers' preference for athleisure clothing. Casual Fridays, that allowed employees to skip their conservative tailored suit and tie or skirt suit for a day, have morphed into the relaxed culture of start-ups. The late Steve Jobs, and former *Apple* CEO, made the casual turtleneck paired with jeans his "corporate" uniform. Mark Zuckerberg, CEO of *Facebook*, has made the headlines more than once for showing up at executive meetings in his hoodie. Greatly influenced by the tech world, employees in a variety of companies favor a type of simple attire that resembles the guise of the next great urban explorer who telecommutes and squeezes in a work out or two during the day. Athleisure is a concept based on new types of synthetic fabrics that are lightweight, allow moisture to evaporate quickly, keep the body warm in glacial temperatures. The selling point is a combination of lifestyle paired with fabric functionality and innovative cuts for performance and movement flexibility. Heavy wools have not been part of this category (Forbes 6 October 2016).

Companies that manufacture urban fashion and athleisure contribute to the production of great numbers of garments perpetuating the evils of "fast fashion," perhaps not as pronounced in urban wear as in athleisure. In the latter category, brands establish business models that lock the consumer in and pressure them to buy several

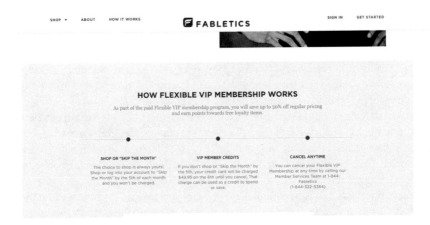

Fig. 9 Fabletics. Website Screenshot (http://www.fabletics.com/)

pieces of clothing per month. *Fabletics* is a good example of that category (Fig. 9). Even without the aggressive business model, marketing in that industry is frequent and intense. As a result, younger consumers, Millennials in particular, habitually change their entire wardrobe every six months or so, a mentality starkly different from previous generations of consumers who would hold on to favorite pieces of clothing for a long time, if not a lifetime. Finally, while a few companies in the athleisure category claim not to harm the environment by producing fabrics out of recycled water bottles and other plastics (as does *Patagonia* for example) (Fig. 10) the reality is that their clothing is not biodegradable. It takes these items anywhere between 20 and 200 years to fully biodegrade.

New modes of professional behavior, greater elasticity of what is or is not accepted as formal wear at the office, aggressive business strategy that leads to overconsumption, deceitful marketing that overemphasizes the reversibility of the harm already done to the planet but does not eliminate the problem of waste, are realities that feed the frenetic fashion engine with the power of a tsunami of global scale. These are the market forces against which the Harris Tweed industry is facing today. The HTA has failed to realize that they should not compete on price against textile manufacturers of machine-made and synthetic products. As a result, independent tweed producers and the other two mills are engaging in a practice that Reid of *Carloway Mill* has been trying to stop: the dumping of Harris Tweed in the market at extremely low prices.

To add insult to injury, the industry seems to be facing enemies domestically. According to news site *The Daily Record*, in 2011, when the new movie of the series "Doctor Who" was produced, the main character who had been traditionally dressed in a Harris Tweed jacket swapped his authentic woolen garment for a "Chinese rip-off that is 20% acrylic" (Merrit 2011). According to the same article, to capitalize on the movie's success, the British Broadcast Corporation (BBC) licensed additional Chinese replica blazers to be sold on the Forbidden

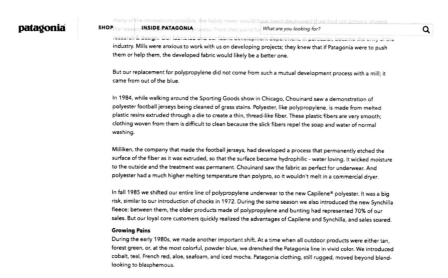

many of the innovations possible, the fabric never would have been developed if we had not actively shaped the research & design. Our fabric lab and our fabric development department, in particular, became the envy of the industry. Mills were anxious to work with us on developing projects; they knew that if Patagonia were to push them or help them, the developed fabric would likely be a better one.

But our replacement for polypropylene did not come from such a mutual development process with a mill; it came from out of the blue.

In 1984, while walking around the Sporting Goods show in Chicago, Chouinard saw a demonstration of polyester football jerseys being cleaned of grass stains. Polyester, like polypropylene, is made from melted plastic resins extruded through a die to create a thin, thread-like fiber. These plastic fibers are very smooth; clothing woven from them is difficult to clean because the slick fibers repel the soap and water of normal washing.

Milliken, the company that made the football jerseys, had developed a process that permanently etched the surface of the fiber as it was extruded, so that the surface became hydrophilic - water loving. It wicked moisture to the outside and the treatment was permanent. Chouinard saw the fabric as perfect for underwear. And polyester had a much higher melting temperature than polypro, so it wouldn't melt in a commercial dryer.

In fall 1985 we shifted our entire line of polypropylene underwear to the new Capilene® polyester. It was a big risk, similar to our introduction of chocks in 1972. During the same season we also introduced the new Synchilla fleece: between them, the older products made of polypropylene and bunting had represented 70% of our sales. But our loyal core customers quickly realized the advantages of Capilene and Synchilla, and sales soared.

Growing Pains

During the early 1980s, we made another important shift. At a time when all outdoor products were either tan, forest green, or, at the most colorful, powder blue, we drenched the Patagonia line in vivid color. We introduced cobalt, teal, French red, aloe, seafoam, and iced mocha. Patagonia clothing, still rugged, moved beyond bland-looking to blasphemous.

Fig. 10 Patagonia. Website Screenshot 'Our History' (http://www.patagonia.com/company-history.html)

Planet website (https://www.fpnyc.com/). This is one of "the largest sellers of comic books, graphic novels, science fiction, toys and associated toys in the world" clearly aiding BBC with its mass market strategy to convert the appeal of the Harris Tweed brand into revenues from the sale a counterfeit product (Chunju 2013).

Finally, another threat has to do with the complexity of the weaving machines. According to Reid, they require a great amount of maintenance hours because the wool can easily clog them. The parts are also based on precise manufacturing and come from Germany. Any costs associated with equipment production directly affect the industry. This has happened several times during the twentieth century and resulted in significant increases of overhead costs for self-employed weavers.

7 Sustainability: The Future of Carloway Mill as a Luxury Enterprise

Due to its financial struggles, the mill was purchased in a Management Buyout (MBO is a formalized type of acquisition where a company's managers acquire part or all of the company from its private owners), led by Annie McDonald, former Head of Operations. McDonald confirmed that the new company maintains Reid's vision that has now been expanded to the following:

Craftsmanship for today and tomorrow's world

We believe that respect for traditional Scottish traditions, perseverance against industry threats, and an open mind will lead us to being a luxury brand within the brand of Harris Tweed.

We believe Carloway Harris Tweed to be synonymous with innovation, superior quality, and adherence to traditional methods of tweed production.

We believe in treating our staff as family and in empowering our weavers in their journey of preserving their craft and transmitting it to the next generation while catalyzing financial prosperity for our region.

We believe in the human element of our work and find pride in our product.

Our vision is to develop the Carloway Harris Tweed brand through the launch of consumer products and to increase consumers' awareness of the superior quality of Carloway products that are fit for contemporary dressing.

We aim to achieve this by complementing our B2B model with a B2C business model.

As discussed in earlier sections, the Carloway brand has associations with a specific place, heritage, and authenticity that define, according to Fiona Anderson, luxury (Anderson 2013). However, the brand has been struggling to operate as a luxury player even though it regularly supplies luxury fashion houses with bespoke tweed (Sweeney 2009).

Today, luxury requires more than Anderson's definition. It shifts our attention to issues of sustainability, specifically as they relate to manufacturing processes and their byproducts. The Harris Tweed industry is by default much closer to William McDonough's and Michael Braungart's concept of "cradle to cradle," namely a system that imitates how nature produces, uses, and recycles in a constant cycle during which the byproduct of each phase becomes nourishment for the next (McDonough and Braungart 2002). The authors' work as they initially conceptualized it for the 2000 World's Fair in Hannover, Germany is a three-pronged manifesto, which developed around the themes of humanity, nature, and technology, and is directly applicable to the discussion of Carloway as a producer of sustainable luxury products. The future of the mill as a luxury enterprise seems promising mainly because the business already respects McDonough's and Braungart's set of principles that were first publicly presented in 1992 and constitute the backbone of their second book, *The Upcycle: Beyond Sustainability* (McDonough and Braungart 2013). These are: 1. Insist on the right of humanity and nature to coexist in a healthy, supportive, diverse, and sustainable condition. 2. Recognize interdependence. 3. Respect relationships between spirit and matter. 4. Accept responsibility for the consequences of design decisions upon human well-being, the viability of natural systems, and their right to coexist. 5. Create safe objects of long-term value. 6. Eliminate the concept of waste. 7. Rely on natural energy flows. 8. Understand the limitations of design. 9. Seek constant improvement by the sharing of knowledge. Principles 1–5 are already at work at Carloway and can be strengthened whereas 6–9 can be applied in tandem with the luxury business model that follows.

According to the author's luxury business model (Fig. 11) (Serdari 2016), the Carloway Mill would be able to reinforce its position as a luxury player by

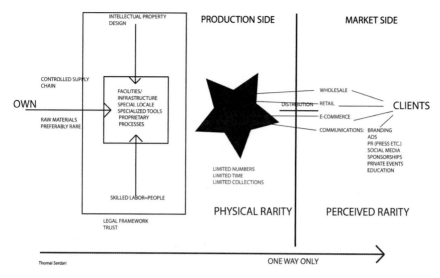

Fig. 11 The Luxury Business Model$^{©}$ Serdari (2016)

revisiting each element of the business model through the lens of sustainability. What follows is the author's interpretation of how sustainability can become a tool for further development of a luxury enterprise and reinforcement of its competitive advantage in the market. This case is aimed for discussion in fashion and business schools so that students can develop additional innovative strategies that will allow the brand to combine sustainable solutions in a traditional textile manufacturing facility.

Checking against the framework of the luxury business model, the following observations apply to the physical and perceived rarity of the *Carloway Harris Tweed*.

7.1   Physical Rarity of Luxury Product/Brand

Supply chain:
As long as local crofters are employed in the breeding of Cheviot and Black Face sheep, the enterprise continues its contribution to sustaining the local economy.
The Australian fleece, while all natural, leaves a substantial carbon footprint during transport to the Outer Hebrides (Black and Farren 2010). It is desirable to investigate whether rather than importing fleece the mill can import a new breed of Australian sheep to be kept by local crofters and what would the implications be for the local microsystem.
Facilities:
The financial investment in restoring the mill's buildings also contributes in sustaining the local economy.

Skilled labor:

As one of the very few employers in the area, the Carloway Mill is responsible for the livelihood of crofters, hand-warpers, weavers, dyeing technicians, mill workers, managers, and administrative personnel. Of these, hand-warpers, weavers and dyeing technicians constitute the Outer Hebrides' living heritage. The learned and tacit knowledge of their respective skill has survived from generation to generation. It must be recorded, documented, archived, and interpreted so that the mill and its product continue adapting to future realities through sustainable and handcrafted means.

Equipment:

Two types of weaving machines are in use and maintained at the mill as mentioned in an earlier section. This is an area of great promise for new product development as *Carloway* has already demonstrated with its new featherweight tweed. In other words, to ensure the longevity of the operation capital investment is needed for research and development so as to catalyze the evolution of the equipment and the way it is used or both and achieve new product development that respects a "cradle to cradle" philosophy.

Manufacturing processes:

While the steps of tweed making are clearly marked by the HTA, the way these are executed offer an area of investigation that does not alter the product or its character. Wool is recognized today as performance fiber and can be made water-repellent and waterproof (Bealer-Rodie 2011). Carloway has an advantage over other companies that are just rediscovering wool and needs to capitalize on its deep knowledge on how to work with wool as an eco-product.

Intellectual property rights:

This is an area that remains exposed. Processes can be patented by the mills and submitted to the HTA for safeguarding. The creative design has always been unprotected, which is why cheap replicas have surfaced in China. The beauty of Harris Tweed has a lot to do with its creative essence, namely the coloring that stems from the land (and its natural flora) as well as the pattern that often reflects the natural environment of the Outer Hebrides.

Limited supply of product:

There is a fixed number of land crofts and pastures, a fixed number of sheep that render their fleece and a fixed number of surviving hand-warpers and weavers. Even when running at full capacity, the Carloway Mill is committed to producing a limited number of yards of tweed.

Limited availability of product:

The Carloway Harris Tweed, a brand within a brand, comes in limited supply and limited availability. The mill produces on contract for its customers and sells directly to consumers at a small shop. This is an area that needs to be redesigned by rethinking the brand's wholesale and retail strategy.

Special product in limited editions:
The mill produces bespoke tweed for established customers. When investments are made in creative development, a variety of bespoke product can be produced in limited editions to enhance the aspect of the product's collectability.

7.2   Perceived Rarity of Luxury Product/Brand

Pricing:
The pricing strategies undertaken by the HTA are not sustainable. Additionally, they undermine the industry's longevity. Reid's initiative to get all three mills' CEOs to meet and agree on stopping the dumping of tweed in the market is the first step in establishing a new pricing strategy. This should communicate the traits that define tweed as luxury (heritage, tradition of craftsmanship, creative artistry, organic materials, tactility, durability, timelessness, and low negative impact on the environment). Additionally, having already differentiated itself within the industry, Carloway Harris Tweed should adopt special pricing to communicate its innovative breakthroughs and creative direction.

Wholesale:
In an effort to combat fast fashion and its negative impact on the planet, wholesale customers should include only luxury fashion houses that do not engage in fast fashion and do not overproduce.

Retail:
The mill should proceed with the development of individual products of the highest artistic and craft quality and sell directly to consumers through carefully branded retail stores. This presents a major capital investment that requires a few years of product development. Creative talent can be invited to apply for residency and absorb the essence of the Outer Hebrides while designing product that resonates with contemporary culture. The Glasgow School of Art would make an ideal partner in nurturing creative talent that can interpret the singularity of the product in contemporary designs. Returning to MacDonough's and Braungart's theory, new product can be designed at Carloway as zero waste fashion utilizing their afore-mentioned principles 6–9.

E-commerce:
An e-commerce branch of the retail operation should be part of the firm's long-term strategy and should be examined separately as Carloway is not yet ready to undertake such an initiative.

Brand building:
Carloway's brand rests on solid foundations but must be developed further to express its singularity. This stems from the firm's innovative presence in a tradi-tional industry and revolves around a fortified strategy in the direction of sustainability.

Advertisements; Public Relations; Social media; Sponsorships; Private Events; and Customer education:
These must be rethought and adjusted once the Carloway brand has been correctly expressed to communicate sustainable luxury.

8 Conclusions

The Harris Tweed industry has been evolving for at least the last eight hundred years. Recognized for the merits of its traditional production methods and quality of final product, it became the main source of livelihood for the inhabitants of the Outer Hebrides in Scotland. Under the auspices of the Harris Tweed Authority (formerly the Harris Tweed Association), the three surviving mills on the islands of Harris and Lewis have been looking for ways to expand their market exposure. Amongst them, the Carloway Mill, under the leadership of three capable entrepreneurs whose passion for Scotland drove them to considerable heights of financial investment, has managed to differentiate itself as a brand within a brand, i.e. the Carloway Mill Harris Tweed. Management chose the strategy of brand differentiation in an effort to combat price dumping in wholesale, buyers' pressure for price cutting, delayed payments that are tied to delivery of product to customers, and most importantly, radically changed consumers' taste. Compelled to modernize its offering and strengthen its bargaining power, management invested in research and development of a new Harris Tweed product that maintains all the qualities of traditional Harris Tweed but is ultra soft and featherweight. In doing so, the Carloway Mill has proven its commitment to the Scottish luxury industry and revealed a new mode of overcoming challenges. This has more to do with reflection, deep understanding of the industry's core and its people, and a clear vision of how to combat the harmful culture of fast fashion as well as deconstruct the myth of up-cycling in textile manufacturing. This brief account of the land, its people, their industry and the business challenges faced by the Carloway Mill showcases that entrepreneurship and sustainability when well aligned can push the concept and practice of luxury production into the twenty-first century. In a business context, reflecting on the luxury industries is a necessity in order to restore the market pace and upgrade our quality of life on the planet.

References

Anderson F (2000) Spinning the ephemeral with the sublime: modernity and landscape in men's fashion 1860–1900. Fashion Theor 9(3):283–394
Anderson F (2013) Tweed. In: Textiles that changed the world. Bloomsbury Academic, London
Bealer-Rodie J (2011) Harris Tweed for a new generation. Textile World 6:46

Berthon P et al (2009) Aesthetics and ephemerality: observing and preserving the luxury brand. Calif Manag Rev 52(1):45–66

Black P, Farren A (2010) The wool industry in Australia and New Zealand. Berg Encyclopedia of world dress and fashion. Australia, New Zealand, and the Pacific Islands, pp 100–106. doi:10. 2752/BEWDF/EDch7017

Black S (2010) Ethical fashion and ecofashion. The berg companion to fashion. Bloomsbury Academic, Oxford

Buxton B (1995) Mingulay: an Island and its people. Birlinn, Endiburgh

Byron R, Hutson J (1999) Local enterprises on the North Atlantic margin: selected contributions to the Fourteenth International Seminar on Marginal Regions. Ashgate, London

Chen J (2017) Why the Supreme and Vuitton Collab Was the Season's Most Brilliant Troll. The Cut. 7 February 2017. http://nymag.com/thecut/2017/02/the-supreme-and-louis-vuitton-collab-was-a-brilliant-troll.html. Accessed 14 Mar 2017

Chunju G (2013) Feature: Scottish Wool Textile Big Brand Seeing Revival Amid protection challenges. Xinhua News Agency-CEIS. 11 April 2013. ProQuest. Accessed 14 Mar 2017

Forbes. The Athleisure Trend Is Here to Stay. 6 October 2016. https://www.forbes.com/sites/greatspeculations/2016/10/06/the-athleisure-trend-is-here-to-stay/#2f34b21c28bd. Accessed 14 Mar 2017

Haswell-Smith H (2004) The Scottish Islands. Canongate, Edinburgh

Herman T (1957) Cottage industries: a reply. Econ Dev Cult Change 5(4):374–375

Kitchener O (1994) Harris Heritage Island Weavers Hand-loom All Harris Tweed in the World. The Record. 18 October 1994: C2. ProQuest. Accessed 14 Mar 2017

Lawson I (2011) From the land comes the cloth. Selvedge 15–25

McDonough W, Braungart M (2002) Cradle to cradle: remaking the way we make things. North Point Press, New York

McDonough W, Braungart M (2013) The upcycle: beyond sustainability—designing for abundance. North Point Press, New York

Merrit M (2011) 'Doctor Who' Swaps Harris Tweed for Chinese Rip-Off as BBC Sells Replicas for £360. Daily Record. 22 May 2011. http://www.dailyrecord.co.uk/entertainment/tv-radio/doctor-who-swaps-harris-tweed-1103613. Accessed 14 Mar 2017

Moisley HA (1961) Harris Tweed: a growing highland industry. Econ Geogr 37(4):353–370

Murray WH (1973) The Islands of Western Scotland: The inner and outer Hebrides. Eyre Methuen, London

Serdari T (2016) Steidl: Printer, Publisher, Alchemist: The Field of Luxury Production in Germany. Luxury: Hist Cult Consumpt 2(2): 33–51

Snedden Economics (2007) Outer Hebrides Tourism Facts and Figures Update. Inverness

Sweeney C (2009) From Paris to Harris: Saving Tweed Is All the Vogue: A Catwalk Show on the Isle of Lewis To Help the Ailing Industry. The Times. 6 March 2009: 25

Thompson F (1969) Harris Tweed: The story of a Hebridean Industry. David & Charles, Newton Abbot, Devon

The New Yorker Archives. archives.newyorker.com

The Vogue Archives. ProQuest

Printed in the United States
By Bookmasters